The COUNTRYMAN SETS FORTH AGAIN

Clarissa and the Countryman
Clarissa and the Countryman Sally Forth
A Sunday Roast
The Game Cookbook
A Greener Life
A Book of Britain
The Countryman Through the Seasons

The
COUNTRYMAN
SETS FORTH AGAIN

Johnny Scott

Quiller

We are indebted to *The Field* for its kind permission to reproduce material originally appearing within its pages.

First published 2023

Quiller
An imprint of Amberley Publishing Ltd

Copyright © Johnny Scott, 2023

The right of Johnny Scott to be identified as the Author of this work has been asserted in accordance with the Copyright, Designs and Patents Act 1988.

ISBN 978 1 8468 9382 7 (hardback)
ISBN 978 1 8468 9383 4 (ebook)

All rights reserved. No part of this book may be reprinted or reproduced or utilised in any form or by any electronic, mechanical or other means, now known or hereafter invented, including photocopying and recording, or in any information storage or retrieval system, without the permission in writing from the Publishers.

British Library Cataloguing in Publication Data.
A catalogue record for this book is available from the British Library.

1 2 3 4 5 6 7 8 9 10

Typesetting by SJmagic DESIGN SERVICES, India.
Printed in the UK.

Quiller

An imprint of Amberley Publishing Ltd
The Hill, Merrywalks, Stroud, GL5 4EP
Tel: 01453 847 800
E-mail: info@quillerbooks.com
Website: www.quillerpublishing.com

Contents

Introduction   7

Spring

Landscape	12
Adders	20
Laughing Frogs	23
Mad March Hares	28
Wise Owl	32
Folklore and Customs	36
The Jethart Ba'	38
Fighting Shrews	40
Charcoal Burners	44
Pigeon Racing	49
Lockdown	53
Bracken	57

Summer

Soft or Hard Grouse?	64
Slippery Eels	70
Golden Gorse	74
Bats	77
Hound Trailing	81
Ravens	87
Reading the Weather	92
Strewing Herbs	96
Ancient Trees	100

Kit	104
Exotica	106
Terriers	113

Autumn

Hanging Game	124
A Mast Year	130
Dryland Huskies	136
Elder	141
Autumn Bounty	145
Wild Fungi	152
Capercaillie	160

Winter

Ferrets	168
Otter	171
Pike	174
Snipe	177
Britain's Alpine Heritage	182
History in a Wall Head	188
Starlings	192
Stoats	197
The Golden Queen of the Woods	201
Coppicing	205
Our Fathers of Old	211
Grey Geese	219
The Language of Field Sports	225
Campanology	231
Sighthounds	237
Yule Log	244
January	250

Introduction

The Countryman Sets Forth Again is the sequel to my previous book published by Quiller, The Countryman Through the Seasons and, like its predecessor, is a compilation of articles I have written over the years for The Field magazine and other publications, containing a mixture of topics including field sports, wildlife, natural history, customs, traditions, and folklore of rural Britain. As before, it is perhaps more in the nature of a scrapbook in which I have recorded vignettes of the odd incidents and episodes which nature has provided to excite my interest, as the wheel of the seasons turns. The sound of a vixen screaming to her mate on a frosty, moonlit winter night; the territorial drumming of cock snipe on a moorland fringe in springtime and the glory of wild flowers in early summer. Badger cubs playing at the mouth of a sett; a hen grouse scuttling through the heather dragging her wing to draw one away from her nest, or the strange outlines on the land that become visible after a fall of snow, indicating some ancient human occupation.

So much of the British countryside has been lost to us during my lifetime, not just through roads, railways, intensive farming and ever-expanding urban sprawl, but because people have lost their connection to the natural world. When I was a child, all children of my age had one thing in common, regardless of background; nature was our principal source of daily entertainment. Urban children learnt about natural history in city parks, railway embankments, churchyards and canal banks, whilst rural ones had the freedom of the countryside. To be outside, whatever the weather, was considered an essential, healthy, beneficial and profitable way for the young to spend their time. These were the philosophies around which the Scout movement, that did such exemplary work in introducing inner-city children to country lore, had been based.

As children we learnt the seasons of birds, animals, reptiles and insects; the ones that hibernated and those that were nocturnal; the predators, the quarry they hunted and the corridors of safety, such as hedgerows, that smaller animals used as habitat, or links to woodland sanctuary. We were taught which plants were edible and those that were poisonous; how to read the weather from the way wildlife responded to differing atmospheric conditions and we knew how to recognise the presence of animals by identifying their tracks. We absorbed our knowledge on a daily basis from our elders – there were many more people working on the land in those days, but the nation as a whole knew more about the natural history of these islands than at any other time before or since, simply through necessity. Rationing lasted from 1940 to 1954 – a staggering fourteen years during which everyone had to forage for nature's seasonal bounty to bolster their diet.

The post-war arcadia of my childhood underwent dramatic changes during the sixties and seventies as government policies of agricultural intensification created a chain reaction of loss across the spectrum of wildlife. Vast quantities of habitat were destroyed in parts of Britain as hedgerows were bulldozed out to create bigger fields, old pasture, heath and downland ploughed and marshes drained under devastating Ministry of Agriculture reclamation schemes aimed at making Britain self-supporting. Herbicides and pesticides destroyed the habitat and food source of many small birds, reptiles and mammals, which in turn impacted on the larger species that depended on them. At the same time, the Forestry Commission embarked on a massive programme of planting quick-growing Sitka spruce conifers. Huge areas of land were planted, much of it in areas of outstanding natural beauty and totally unsuited to growing trees, the Highlands of Scotland, for example, or the uplands of Wales and northern England – Kielder Forest alone sprawls over 160 square miles of the Northumbrian hills. Rural communities disappeared; acres of ancient natural woodland were engulfed and in a matter of twenty years, these plantings had become vast black blocks of sterile woodland.

As machinery increasingly replaced manpower, 50 per cent of the agricultural workforce left the land to seek alternative employment in towns. The self-supporting infrastructure of villages became eroded, many of the old rural crafts died out and much country lore, handed down from generation to generation, was lost. As agricultural reclamations destroyed the hedgerows and small broadleaved woodland, they took with them the urban tradition of coming out into the countryside every autumn to pick nuts and berries.

Technology has now enabled more and more people with urban-based employment to live in rural areas and to reap the benefits for themselves and their families from living in the countryside and yet, so many of them feel ill at ease in the midst of their natural heritage through lack of knowledge. A combination of bureaucracy, ill-advised agricultural policies, ignorance and prejudice have combined to obliterate so much of the precious culture of our Sceptred Isle. Gradually, the much vaunted urban–rural divide became established, fuelled by field sports becoming a political football as socialist politicians encouraged lobbying by commercially motivated single issue animal rights activists, forcing genuine country people into becoming an ethnic minority.

It is all still there though. The romance, antiquity and beauty, but our rural culture, heritage and traditions; the catalyst that binds many communities together and provides them with the folklore that defines their regional identity has become very fragile and sadly, there are those motivated by personal gain, who seek to destroy it for ever.

The wonder of the world; The beauty and the power,
The shapes of things; Their colours, lights and shades,
These I saw; Look ye also whilst life lasts.

Spring

Landscape

The British landscape is an extraordinary creation; immensely ancient and full of enchanting surprises which open little windows of our history. I cannot believe that any other country has such a diversity of interest packed into a smaller space. It is impossible to go from one parish to another without coming across some arresting reminder of the country's past, each with a story to tell – an Iron Age fort, a strangely corrugated field, a ruin, a folly, a venerable tree, a stone circle, castle, sunken lane, ancient bridlepath, right of way, old stone farm building or simply an isolated patch of nettles, indicating that humans had once settled in the immediate area. Every day on my farm here in a remote part of the Scottish Borders, I walk past the physical memorials to previous occupiers of this land going back dozens of centuries. On a bank above the Whitrope Water is a boggy area of ground called Buckstone Moss, named after the Buck Stone, a Neolithic megalith erected perhaps 3,500 years ago by dreamy prehistoric pastoralists. There are the visible remains of the earth banks that surrounded the little fields attached to the Iron Age fort on a hill called the Lady's Knowe. Below these lie the Lady's Well, a freshwater spring revered by the Celts long before the nearby chapel, part of the Hermitage Castle, was dedicated to St Mary.

Between the Lady's Well and the ruins of St Mary's Chapel are a jumble of mounds and earth banks assumed to be the remains of the motte-and-bailey castle built by Sir Nicholas de Soules, Lord of Liddesdale, in 1240. Further on, beside the Hermitage Water, on a bank above a deep pool is an oblong hump, reputedly the grave of Sir Richard Knout, Sheriff of Northumberland, who was killed by retainers of the de Soules family in 1290 when they rolled him, in his armour, 'into the frothy linn'. Then there is the grim awesome ruin of the Hermitage Castle, 'the guardhouse of the bloodiest valley in Britain', where, in 1566, Mary, Queen of Scots had the infamous meeting with her lover, James Hepburn, Earl of Bothwell. Back in the body of the farm, a great wall of boulders, known as the White Dyke, runs across the middle of Hermitage Hill, said to be part of the deer 'haye' or funnel into which deer from the castle deer park were driven and slaughtered.

There are more stone walls or dykes built in the eighteenth century during the Acts of Enclosure, when gangs of Irish labourers built mile upon mile of walling across Scotland and Northern England. At much the same time, drainers dug open drains all over the hill to improve the quality of the grazing and built 'cundies' (conduits) to carry water from one of the hill burns to power the water mill at the steading. An old drove road runs down the side of the farm's northern boundary through an area known as the Mount; at the bottom are the ruins of an old toll house and the earth banks of Mount Park, where cattle from all over south-west Scotland rested for the night on their long journeys to the trysts in the north of England. The 'old' steading, a handsome range of slate-roofed stone buildings – cattle byres, cart sheds, granary and stabling – was built in 1835; the 'new' steading, a hideous open-span erection of steel girders, asbestos and concrete, was put up in the 1970s, when the government was offering subsidies for new farm buildings during a drive to increase agricultural output. I mention all this in detail because my farm only covers 600 hectares and although having a castle on the doorstep adds a certain amount of added historical interest, the visible traces of preceding generations are similar to those of all other farms in the country.

Virtually every corner of the British Isles, from the tip of Cornwall to remotest Hebridean Island, has been owned and tilled, cropped and grazed for at least 7,000 years. For all its wonderful areas of remote, rugged and natural beauty – the Cumbrian Fells, the Cheviot Hills, the savage grandeur of the Highlands or the moorland of the West Country – Britain is the least wild of any country on the planet. It has been estimated there is not a yard of land that has not been utilised by someone since the arrival of Neolithic man, and the landscape we love and admire is entirely man-made. The rolling heather-clad hills of Scotland are most certainly man-made – even the Broads, the stunning network of lakes and rivers covering 300 sq km of Norfolk and Suffolk. Until the 1960s, when the botanist and stratigrapher Dr Joyce Lambert proved otherwise, this vast wetland area was believed to be a natural formation. In fact, they are the flooded excavations created by centuries of peat extraction. The Romans first exploited the rich peat beds of this flat, treeless region for fuel, and in the Middle Ages the local monasteries began to excavate the peat as a lucrative business, selling fuel to Norwich, Great Yarmouth and the surrounding area. Norwich Cathedral, one of the most stunning ecclesiastical buildings in Britain, was built with money from 320,000 tonnes of peat dug out of the Benedictine lands every year, until the sea levels began to rise and the pits flooded.

During this incredible longevity of occupancy we have developed a passion for our countryside, a bond and an affinity with the land that is uniquely British. This love affair has been expressed throughout history by an almost obsessive desire to draw attention to the landscape by affectionately adding what was considered at the time to be an improvement to nature's already superlative offering. Britain is covered in decorated summits, follies, woodland plantings, individual trees, artificial lakes and monuments, all carefully sited to improve the vista and all constructed as a statement of gratitude.

Our Bronze Age and Iron Age ancestors were among the most diligent of landscape enhancers, compulsively building henges, erecting megaliths and carving hill figures where the colour of the chalk or limestone substrata would show up in contrast with the green of the surrounding sward. Undoubtedly the most famous of these is the White Horse of Uffington, high on an escarpment of the Berkshire Downs below Whitehorse Hill, a mile and a half south of the village of Uffington, looking out over the Vale of the White Horse.

For a piece of artwork which optically stimulated luminescence dating has proved to be 3,000 years old, the highly stylised curving design is extraordinarily contemporary. It was either the late Bronze Age or early Iron Age occupants of the adjacent Uffington Castle hill fort who devoted the immense amount of time, organisation and effort required to carve the 110-metre creature into the hillside and, despite endless hypothesis, no one really knows why. From my perspective, you only have to look at it for an explanation: the horse is a thing of beauty, young, sleek and vibrant, lunging forward with neck arched and forefeet raised, a picture of health and vitality. The carving was deliberately constructed just below the summit where it would be visible to other hilltop settlements and the horse triumphantly shouts a message from his tribe across the wooded valleys: 'Look at me.' The horse rejoices, 'Am I not magnificent? See how beautiful and fertile my hill is.'

Unless the substrata was regularly kept exposed, a hill carving would disappear back into the ground within a decade and there will have been hundreds of them dotted around the uplands which are now lost to us. The two Plymouth Hoe Giants, visible until the early seventeenth century, are an example, or the Firle Corn, a nearly lost hill figure on Firle Beacon, in Sussex, now looking more like a small ear of corn or a strange weapon than a human figure, whose existence can only be seen by infrared photography. What is so remarkable about the Uffington Horse is that for over thirty centuries whenever the turf looked like growing over it the local people have always cleared it away. Long after the original architects had passed on and whatever religious, totemic or cultural significance attached to the carving had been forgotten, successive generations have preserved the carving through all vicissitudes, simply because they liked having the horse on their hill and felt it looked better with it, rather than without it.

Some hill figures are of dubious provenance. There are no historical records of the priapismic Cerne Abbas Giant before 1694 and there is considerable evidence to suggest that it was created on the instructions of the landowner, Lord Holles, as a parody of Cromwell. Others have been resurrected by nineteenth- and twentieth-century archaeologists – whose enthusiasm has almost certainly changed the original outlines. The Long Man of Wilmington dominating a broad sweep of the South Downs near Eastbourne is an example. The origins of this 70-metre-high figure have been the subject of endless debate, ranging from a heretical image carved by a secret occult sect of the monks of Wilmington Priory during the Middle Ages; a Celtic sun god opening the dawn portals and letting the ripening light of spring flood through; a Roman standard bearer, or a deeply symbolic prehistoric fertility symbol. Unfortunately, many of the original features were lost in 1874, when the outline was altered to make it appear more impressive, but I have no doubt that the Long Man was made by the late Bronze or early Iron Age tribesmen who occupied a substantial settlement on the summit of Windover Hill immediately above it.

Why did the ancients carve a giant man there? I believe, as with the White Horse, they were broadcasting pride of ownership of that particular hill settlement. One thing is certain, the lovely curvature of the Downs and the uniform, slightly convex slope between the two almost identical spurs on which the Long Man has been carved would pass completely unnoticed if he wasn't there. All the hill carvings, the few ancient ones which have survived or been resurrected and the many that were created in the nineteenth century during the great era of naturalist landscape design, draw the eye to a pleasing feature of landscape.

The exertion that went into digging out hill carvings pales into insignificance when compared with some of the other creations that display an extraordinary commitment of time and effort for no apparent purpose. Silbury Hill near Avebury, in Wiltshire, is the tallest, prehistoric, human-made mound in Europe and one of the largest in the world — similar in size to some of the smaller Egyptian pyramids of the Giza Necropolis. Composed mainly of chalk and clay excavated from the surrounding area, the mound stands 40 metres high and covers about 2 hectares. It is an exhibition of immense technical skill and prolonged control over labour and resources. Archaeologists calculate that Silbury Hill was built nearly 5,000 years ago and took 18 million man-hours, or 5,000 men working flat out for fifteen years to deposit and shape 250,000 cubic metres of material. This incredible structure contains absolutely nothing; no burial chamber of a great tribal chief and not one iota of treasure. A huge disappointment to the first Duke of Northumberland who employed an army of Cornish miners to burrow their way through the hill in 1766, convinced they would find him some loot. There is no explanation why anyone should want to build Silbury Hill, apart from the indisputable fact that it looks jolly impressive in the middle of an otherwise flat piece of ground.

Equally peculiar are the inexplicable earthworks known variously as black-dykes, devil's dykes or Grim's dykes, found from the south of England right up into southern Scotland and as far north as Shetland. These consist of a ditch and mound of varying dimensions which follow a winding course across country, often traceable for miles. The great trench and mound of the Devil's Dyke in Cambridgeshire and the long line of Offa's Dyke on the Welsh Marches are two of the most well known. The Devil's Dyke runs for 12 km from the flat farmland of Reach, past Newmarket to the wooded hills around Woodditton, periodically reaching a height of 11 metres. Offa's Dyke is the massive 200-km linear earthwork, 20 metres wide and about 3 metres high, which roughly follows part of the current border between England and Wales. There are several other remains of earth banking: Grim's Ditch in Harrow; the Black Ditches at Cavenham in Suffolk; the Brent, Bran and Fleam Ditches in Cambridge, and Woden's Dyke in Wiltshire. In southern Scotland we have the Catrail, which meanders 22 km from Roberts Linn, just up from the farm, to Hoscote Burn in south-western Roxburghshire. The 8-km Picts Work Ditch, from Linglie Hill to Mossilee, near Galashiels, and the Celtic Dyke in Nithsdale, Dumfriesshire, runs for about 27 km parallel with the River Nith between New Cumnock and Enterkinfoot.

Scottish 'black dykes' are small compared to the others, being about 2.5 metres at the base. Most of these earthworks appear to have been constructed by Bronze Age and Iron Age people, with some in the early Anglo-Saxon period and all, even Offa's Dyke, share one thing in common: for all the labour and energy that must have gone into building them, they serve no recognisable function. They are demonstrably not defensive; in most cases they are so short that an enemy would simply nip round the sides or, in the case of Offa's Dyke, it would be impossible to man the entire length effectively. They are obviously not boundaries and a theory popular among nineteenth-century Scottish historians, that they were built to hinder neighbouring tribes escaping with stolen livestock, was quickly discredited. The sort of semi-wild farm animals that were around in those days would easily have been driven through the wide ditch and up the slope of the earthwork.

I find it absolutely delightful that these ancient earthworks have completely stumped the theorists and not even the silliest neo-pagan can claim them as some sort of fertility symbol. So why were they built? In the absence of any other explanation, I presume the motive was similar to that which gave us Silbury Hill; someone must simply have woken up one morning and thought a big earth dyke in this or that location would improve the look of the landscape.

Adders

A little April sunshine and the moorland above the farmhouse throbs with the birdsong of migrants from their coastal winterings who nest every year in the uplands: oystercatchers, curlews, lapwings, snipe, redshanks, greenshanks, golden plover, skylarks, merlins, short-eared owls and the occasional dotterel. A glorious cacophony of mating calls, heralding the start of spring – as is the appearance of the adders I see every year, basking on a bare patch of ground among old heather on a south-facing bank above their hibernation site, the heap of stone from an old sheep stell that collapsed long ago. *Vipera berus,* Britain's only venomous reptile, are short stocky creatures with disproportionately small heads, easily identifiable by the dark zigzag pattern along the length of their back. The male is between 60 and 70 cm long and varying colours of grey and khaki, whilst the larger, stouter female is more brown or red. In both cases the belly is grey and the throat, cream. Adders have the widest distribution of any snake and are found throughout Europe and Scandinavia, across Russia and Asia to northern China and are the only species found inside the Arctic Circle. They occur in a range of habitats from heathland, woodland glades and moors to grassy cliff tops, sand dunes and railway embankments. Their principal requirement seems to be an open, dry, sunny environment free of urban populations, with enough rough cover to escape to when disturbed.

Adders seek hibernation sites when the glass drops in October, with males emerging in early spring several weeks before females, to absorb warmth and vitamin D from sunlight as their sperm develops. They do not feed during this period and continue to live off remaining body fats stored the previous summer. In due course, as the females start appearing they become more active; last year's dusty skin is shed and the male, gleaming brightly, sets off following a female's scent trail with his tongue, which acts as the olfactory organ. Once a female has been located, the male writhes over and round her, flicking his tongue along her body and tapping her with his head, becoming almost convulsively frantic in his attentions as she responds and releases pheromones. If another male appears, the first will defend his position aggressively and an 'adder dance' ensues taking the form of a wrestling match until one concedes defeat. To assist copulation, adders have been provided with a bifurcated hemipenes covered in spines and once male and female have joined, they may remain locked for up to an hour; if disturbed, the female glides away dragging the male behind her.

Sexual maturity is reached by males at three to four years and females, a season later. Female adders who have bred are generally desperately undernourished as they enter hibernation and only reproduce at most, every other year. Adders give birth to live young and with gestation lasting around twelve weeks, the female needs to feed hard as soon as mating is over to build up fat before advancing pregnancy reduces her ability to hunt. Between five to fifteen brick-red young, 20 cm long are born fully active towards the end of August and into September, receiving no parental attention and leaving their mother within thirty-six hours. Prey consists of small mammals – shrews, voles and field mice, lizards, frogs, toads and during the nesting season, fledglings of ground-nesting birds. Adders hunt using their tongue to follow scent and heat sensors located on their slightly upturned snout which detects the prey's body warmth. Quarry is killed with a bite from hollow, hinged fangs through which venom is injected from glands in the upper jaw. These fold back and the wide opening upper and lower jaws move independently, allowing adders to swallow victims whole, whilst enzymes in the digestive tract break down all bones and tissue except fur.

Although poisonous, adders are not aggressive and their danger is largely exaggerated. They are more anxious to avoid people than to attack and usually slither into cover when they detect the ground vibration of an approaching human. Bites happen through accident or stupidity and are rarely fatal – the last death was in 1975 – the effect varies from localised discomfort and swelling, to vomiting and diarrhoea, although dogs are particularly at risk in the spring when the haemotoxic venom is more potent. Until well into the twentieth century, ointments made from adder fat were the universal cure-all for bruises, rheumatism, deafness and snake bite, with adder catchers in various parts of the country making a decent living catching and selling adders to quack doctors. Perhaps the most famous snake catcher was Harry Mills, who died in 1905 and is reputed to have caught over 20,000 snakes in the New Forest during his life. Since 1981, adders have been protected under the Wildlife and Countryside Act from being killed, injured or sold.

Adders have few predators. A hungry brock will take the odd one; a fox with nothing better to do might torment an adder until it is exhausted and kill it. Hedgehogs occasionally eat them and basking adders run the risk of being spotted by buzzards. Although not threatened with extinction in Britain or considered a priority species under the UK Biodiversity Action Plan, adders are endangered in certain areas through habitat loss to agricultural reclamation, motorways and urban spread. Assessing the adder population is now of considerable importance and the Herpetological Conservation Trust would be enormously grateful for information on adder sightings this spring, so keep your eyes peeled.

Laughing Frogs

The Cooling Marshes, on the Hoo Peninsula, was an area I knew well in the 1970s. Part of the North Kent marshes that run beside the Thames from Shorne to Allhallows at the mouth of the estuary, they are an isolated sanctuary in a heavily urbanised area surrounded by industrialisation. All marshes are atmospheric but these ancient grazings are more so than any others I know. Iron Age settlers, the Belgae, Romans and Saxons all left their mark. During the eighteenth and nineteenth centuries the tidal creeks were the scene of intense smuggling activity, with contraband carried to the sinister Shades House, from Egypt Bay, a sheltered anchorage named after African slavers and their cargoes of 'Egyptians'. These marshes were the inspiration for Dickens' *Great Expectations*, but above all, they have always been a haven for a multitude of wildfowl, a natural bounty harvested for centuries to supply the demand in London, as names like Decoy Farm and Decoy Fleet suggest.

I came back to this eerie part of the world for the first time in twenty-five years the summer before last, to see an area of marsh that had been purchased by the Kent Wildfowling and Conservation Association with support of grant aid from the Nature Conservancy Council. Following the track past the ruins of Cooling Castle, out towards the sea wall to meet Allen Jarrett, the KWCA president and Phil Eliot, who manages this part of their landholding, at the famous Cooling Black Hut, a wildfowlers' refuge built from driftwood under the sea wall.

Walking across the marshes to look at one of the 'scrapes', shallow depressions which when flooded, provide KWCA members with spectacular duck flighting, I could hear above the sound of teeming flocks of waterfowl and waders a curious, incessant quacking from a big ditch behind the mounds of old Roman pottery workings. It was an odd sound, that stopped just short of recognisable. 'Widgeon?' I enquired. 'That's what I thought, the first time I heard them,' Phil told me, 'but it's not. It's those Continental frogs. There are thousands of them, all over the marshes.'

Almost immediately I saw one, sunning itself among the sea club rushes on the edge of a dyke. About the size of a common toad, with a pointed snout, light-green head and olive body speckled with black, warty protuberances, the frog gazed at me defiantly through golden eyes before leaping sideways, to land with a plop in the water. This started a chain reaction down the length of the dyke as more basking frogs, sensing a predator, took to the water. With identification, their raucous quacking, which had at first seemed localised, now became an irritating background sound right across the area. A foreign reptilian intrusion on the familiar beauty of marshland music.

Rana ridibunda, the 'laughing frog', so called because its duck-like mating call resembles a chuckle, are the largest frog in Europe – Phil has seen them with a body as big as 13 cm long – has a distribution among European river valleys, ranging from Denmark to the southern Balkans and across Spain, Portugal and the south of France. More commonly called marsh frogs, they were unknown in Britain before the mid-1930s and the story of their introduction and subsequent colonisation is, like most invasive species, a pretty quixotic one.

In 1935, the playwright, and author, E. P. Smith, acquired twelve marsh frogs from University College, London, which had been imported for scientific research from Debrecen in Eastern Hungary. These were released into the pond in Smith's garden at Stone-in-Oxney beside the old Royal Military Canal, overlooking the lovely Walland and Romney marshes. It has been suggested that Smith wanted the frogs to remove mosquito larvae from his pond and that the idea, in those pre-DDT days, was inspired by publicity surrounding the recent introduction of giant marine toads to control grey-backed beetles in the Queensland cane fields.

The frogs, who have a voracious appetite, will eat virtually anything they can get their mouth round, including all invertebrates, small crustaceans and fish, adult newts, other frogs and the occasional mouse or fledgling. They made no impact on the mosquitoes and, preferring deeper water than was on offer in Smith's pond, very soon decamped. The network of drainage ditches, dykes and waterways that criss-cross the marshes proved an acceptable habitat and the frogs began to breed. Female marsh frogs are very prolific and can lay between 4,000 and 12,000 eggs, depending on body weight. Not surprisingly, within three years they had colonised an area of Romney Marsh extending to 46 sq km. E. P. Smith, who took a very proprietorial interest in the frogs' progress, as fellow members of the Garrick Club knew to their cost, was apparently delighted. The inhabitants of the marshland villages, considerably less so.

During the breeding season which starts in May, male marsh frogs become extremely vocal, inflating a grey, pea-sized sac on either side of the mouth to emit their loud, exuberant love song. The noise, which can carry for miles and is described succinctly by Phil as 'deafening', reaches a crescendo through June and continues sporadically into September, when the weather cools and the frogs start thinking about hibernating. Unlike our common frog which mates in February and March, considerately piping down after dusk, marsh frogs become even more vocal during the night. Such was the noise disturbance that the local MP was soon being lobbied by furious residents, demented through lack of sleep, to raise a question in the House of Commons.

Unfortunately, political interest in marsh frogs was overtaken by world events and they continued to spread unimpeded across the marshes throughout the war years.

By 1960, the whole of Romney, Walland, Denge and Winchelsea Marshes, an area of 160 sq km, was thoroughly colonised and E. P. Smith had prudently moved to another part of the country. They were on the Isle of Sheppey by the early seventies and Phil mistook them for widgeon on Stoke Saltings, at the mouth of the Medway, early in the season of 1979. He first saw them out on Cooling Marshes nine years ago. Meanwhile, marsh frogs were becoming established by human agency elsewhere. A farm labourer near the Lewes area brought some frogs back from the Romney Marshes in 1974, which subsequently escaped into the Lewes Brooks and have spread across marshland on either side of the River Ouse, south to Newhaven and north to Barcombe. There is another population in the vicinity of Heathrow, reputedly turned into the wild by a soft-hearted customs official and there have been further releases on the Somerset Levels and further north, on the marshes off the Humber.

A number of factors favour the increase and spread of marsh frogs. Their ideal habitat is warm wet lowlands, with a preponderance of slow-moving deep water and we have no shortage of that. There is a constant food supply and our climate is getting warmer, giving marsh frogs the opportunity of an extended range of distribution. Being primarily aquatic and having toxic skin secretions that make them and their tadpoles taste revolting, marsh frogs have few predators other than herons, little egrets and grass snakes. When basking on the edge of water they are constantly alert and dive for safety at the first hint of terrestrial danger. Pollution and habitat destruction through agricultural reclamation were threats during the eighties, but conservation awareness has removed this in both cases.

Some wildlife organisations insist that marsh frogs occupy an ecological niche that poses no threat or habitat competition to our native amphibians. I am certain they are wrong. Marsh frogs are predators and their increasing numbers must impact, sooner or later, on existing biodiversities. The fact that Phil has not seen a common frog on the inland marshes at Cooling for the last eight years, is conclusive in my view. The history of damage to the ecology by introduced exotic species looks about to be repeated. Except, that marsh frogs are edible and France imports 3,500 tonnes of frogs' legs per annum from Eastern Europe. It's just a thought.

Mad March Hares

The late train from London gets into Newcastle at midnight and, barring mishaps, the drive home takes another hour and a half. The last part of the journey is on a tiny back road that winds its way along the edge of the moors, with open hill on one side and fields of the marginal upland on the other. Coming back under a brilliant full moon the other night, with the windows open and lights off, listening for the first sound of spring – the exultant yapping of oystercatchers, I rounded a corner by a windbreak of beech and larch trees to find a company of six brown hares in the middle of the road. Two were engaged in a classic 'mad March hare' pose – nose to nose like sumo wrestlers, the others crouched, watching in eager anticipation.

Nor did the audience have long to wait. The protagonists sprang upright and engaged briefly in an exchange of blows with their forefeet, turned and tore down the road shoulder to shoulder. They stopped abruptly and jumped over each other kicking with their hind legs, boxed again and rushed back towards me. All six swirled up and down the road as the two bucks leapt and pranced. They gambolled, chased, kicked and boxed in a frenzy of energy before the whole lot swarmed over a dry-stone wall and careered off down the side of a field. Depending on the weather, brown hares breed at any time of the year and the exuberant lunacy that afflicts them during March must have as much to do with their reaction to the start of spring growth and certainty of summer's warmth, as any urge to rut.

There are three species of hare in the British Isles, the blue or mountain hare which is indigenous, turns from chocolate brown to white in winter and lives on high moorland. The larger, lowland brown hare which was introduced and the Irish hare, a recognised species in its own right, exclusive to Ireland. These combine some of the characteristics of both, being larger than a mountain hare with a coat that turns partially white in winter. Blue hares can be found on high ground across northern Europe and the Arctic regions. Brown hares have a natural distribution that extends across the whole of Europe to central Asia and were introduced during the 1800s for sport to North and South America, Australia and New Zealand.

No other animal is more surrounded by myths and legends or commands such respect among sportsmen than the hare. Most of it attached to brown hares, partly because they are bigger, faster, more extreme in every way and infinitely better eating than their mountain cousins, but also I suspect, because they played such an important role in Celtic religion. Oestre, the pagan goddess of dawn, fertility and rebirth, had a hare as her favourite animal, light bearer and attendant spirit. They were also the symbol of Andrasta, patron goddess of the Anglo-Saxon Iceni, to whom hares were periodically sacrificed, which would seem to contradict the accepted belief that the Romans were the first to introduce brown hares to Britain. Paradoxically, the wildest of all animals can be easily tamed, if caught young enough and the Celtic ruling classes liked to keep them in their houses as a sort of living connection to the gods. Boadicea is reputed to have careered into battle with the family pet stuffed inside her blouse.

Solitary and nocturnal, brown hares are the fastest and most agile European animal, capable of speeds of 65 kph. They can turn on a sixpence in full flight and jump over 6 metres with ease. This in itself was enough to command respect from early people but their behaviour, which sometimes appears almost humanly irrational and a hideously childlike scream when caught or injured, convinced them that hares were more than simply animals. As surface dwellers, hares rely on speed, eyesight and acute hearing for protection, and have developed elaborate defensive tactics to disguise their scent from predators. Hares feed at dusk and dawn, lying up by day and eating the soft faeces from the previous nights' feed. When leaving or approaching their form, they double back and forth, making sudden leaps and 90-degree turns to break up the scent line.

For centuries, there was a common belief that hares were hermaphrodite and both sexes bred. This was disproved in the nineteenth century, but what is almost unique, is their ability to be pregnant and conceive at the same time. Does can have four litters of between two and four leverets a year and unlike most young animals that are born blind, bald and defenceless, these have fur and can see immediately. Once the birthing process is over, the doe scrupulously cleans each one and moves them to separate hiding places where they wait, silent and immobile for their daily feed. About an hour after sunset, breaking her scent trail in the usual way, the doe suckles each leveret in turn. Once fed, she rolls the little animal on its back and stimulates the urinary area with her tongue, ingesting any discharge to ensure it remains scent free for the next twenty-four hours before moving them to new sites. Leverets become independent of their mothers after a month.

Under Norman Forest Law, hares were elevated to be one of the five noble beasts of venery, joining the hart, hind, boar and wolf. The Normans were great hound men and loved hunting. Deer, boar and wolves provided speed and drama, but for the complicated science of hound work there is still nothing to beat a hare. They remained fiercely protected by successive game laws, until suddenly demoted to vermin by the Ground Game Act of 1880. Nearly 200 years of agricultural improvements and the small field mixed farming policy of the time, providing an ideal habitat and food range, had led to a dramatic increase in hare numbers – a conservative estimate of the winter population exceeded 4 million. With the right to take game restricted to landlords, tenant farmers struggling to survive decades of agricultural depression lobbied Parliament to be allowed to protect their crops at any time and by whatever means. Hare numbers dropped so alarmingly in the decade following the Act, that a feeble effort to stop the decline, the Hare Preservation Act – which simply banned the sale of hares between March and July, was passed in 1892. It has been left to the sporting community to give hares the courtesy of a closed season.

Brown hares suffered another blow during the 1970s and 1980s from modern farming. The traditional crop rotation of roots, grass leys and cereals which had previously provided hares with a consistent food source was replaced by arable monoculture. Pesticides removed weed plant varieties essential to their diet and silage, cut so much earlier than hay, destroyed breeding habitat and killed leverets born just before cutting. The population dropped to 400,000 and the only areas hares could be seen in any number were those where landlords, who supported coursing under National Coursing Club rules, farmed for hare conservation. In the last twenty years, EU arable set aside rules subsidising farmers for land taken out of production and the more recent provision, allowing this allocation to be used to create broad headlands, has done much to replace lost habitat and brown hare numbers have more than doubled. The greatest risk to brown hares in this day and age, is that their conservation has been taken out of the hands of those who love and respect them by the Hunting Act.

Wise Owl

For many years, a succession of barn owls have nested in the granary and hayloft of the derelict farm buildings behind the house. It is an ideal, safe habitat for them; most of the slate roof is still weatherproof and the flooring so riddled with woodworm that there is little risk of human disturbance. As light begins to fade on a winter's afternoon one, then the other, floats from the doorway above the cart shed's arched entrance and drifts silently away to hunt over the moor or woodland, announcing each success with a shriek. When they return, I often hear them long into the night, chattering to each other in a series of weird catarrhal hisses. The same sound can also indicate the presence of hungry chicks during the summer.

Back in April, during the avian influenza alarm, there was a possibility that free-range chickens would have to be housed. I have a flock of Oxford gamefowl, which normally lead a feral existence around the steading and adjacent wood. To conform with anticipated regulations, I rigged netting across a lean-to. The mesh has not been taken down and the other evening I was horrified to find an owl hanging upside down in it by a leg.

Mercifully, it must have just begun the evening's hunt and not been trapped for long but freeing it without help was impossible. While waiting for this to arrive, I held the wings close to its body. To pass the time, I stroked its head feathers and the owl adopted a surrender posture typical of wild animals. The rigid body relaxed, outstretched talons closed, furious black eyes became concealed as yellow eyelids drooped and it appeared to go to sleep, giving me an opportunity to study this beautiful bird at close quarters.

A barn owl's plumage, white underparts and silvery, mottled buff head and neck with the exquisite texture of silk allows it to hunt small mammals with acute hearing – field voles, mice and shrews – in absolute silence. In fact, so essential is this camouflage that increased sound created by wet feathers precludes hunting in rain. The white oval face is bisected by a central ridge of tightly packed, small feathers, which deflects sound to asymmetrically positioned ears, enabling owls to detect the faintest movement and hunt in the dark with absolute accuracy. Both sexes are about 34 cm long with a wingspan of around 90 cm. They appear to be stocky, but surprisingly the body felt small through the compressed feathers and the talons looked disproportionately large.

When help arrived the owl came to, angling its head towards the sound of scissors snipping. Once freed, it drew both legs up to its body, demonstrating that there was no injury, and when thrown gently airborne it flew slowly in expanding circles over us before dipping low to pick up speed and move off to resume hunting. Later, I heard the familiar, protracted hissing of two owls.

I was intrigued to know whether there were young in the loft. Barn owls have an extended breeding season, starting in February, but subsequent brood sizes are dependent on weather and food availability. Adults require a diet of up to five small mammals a day and owlets quickly reach the same intake. The hen is fed by the cock for nearly a month while she incubates an average clutch of four eggs. The chicks do not fly until they are eight to ten weeks old. Last winter was long and I heard reports that many owls were in poor condition this spring, suggesting barren nests or late broods.

I have not ventured into the loft since part of the floor collapsed a couple of years ago but curiosity, coupled with the excuse to survey the building's condition, overcame prudence. The loss of nesting sites due to demolition or conversion of redundant farm buildings in the seventies and eighties contributed to the decline of barn owls and I consider myself lucky that these still stand. The 35-metre loft is a dark, eerie place festooned with cobwebs. Swallows nest here in the summer and a colony of pipistrelle bats lives behind the plaster and lath walls. Scattered around are mouldering relics from days when the stalls below had carthorses: collars with straw stuffing spilling out, bits of harness, rusting shaft tugs and stoneware bottles. Some of the roof supports had crumbled since I was last up here and more of the floor seemed to be missing.

I was on the point of retreating down the ladder when there was a flash of white and a 'splat'; a barn owl floated towards me, passed without a sound and landed on a beam. I tiptoed cautiously in that direction. Soon piles of fresh, shiny, black owl pellets became visible, surrounded by splashes of white excreta and the detritus of older pellets. Made of the indigestible parts of an owl's prey and generally regurgitated before they set out to hunt, pellets give a fascinating insight to owls' diets. I broke open one that was a tightly packed wodge of fur and little bones – jaws, vertebrae, ribs and tibiae – of at least three small rodents as well as the wings of dor beetles. When hunting hard, a barn owl's prey is extended to include frogs, moles, bats and small birds.

A nest is usually made from a pile of pellets that have been stamped into a hollow to contain the eggs. I peered about in the hope of seeing one. There seemed to be nothing in sight until I glanced up where the roof joins the wall head. Years ago, clay drainage pipes had been stored here and an owl was watching me over the top of them. Detection broke its nerve and it flew off to perch on a beam. From the cavity behind the pipes, I heard a faint hissing and the clicking of tiny beaks. Curiosity satisfied, my return journey seemed to take an age. Every step was an agony of expected disaster. Safety of the ladder was finally reached with the conviction that nothing would induce me back up it again. That evening, I telephoned a falconer friend of mine who educates inner-city children about the countryside and birds of prey. I told him about the owl in the netting, my loft adventure, the pellets and the nest. 'I use owl pellets in my demonstrations with kids,' he told me. 'Could you get me some?'

Folklore and Customs

No country can have as many folk customs, festivals, feasts, privileges and traditions than the British Isles. Other nations use their folklore to define a single patriotic identity; ours define regional communities. Each county, every city, town, village and hamlet has its own unique variation. There are festivals with pagan origins which were subsequently adopted by the Church for every month of the year; strange ceremonies, such as Beating the Boundaries or Riding the Commons; wakes fairs and winter fire festivals. Odd sports, games and dances which have remained unchanged since the early Middle Ages; obscure medieval, civic and manorial courts.

Some customs, such as the ancient Doles and Charities, still have a practical application. Others are so old that they have long since lost their original point and are still practised simply because they have become embedded in the calendar of the locality. Blowing the wakeman's curfew horn every evening in Ripon, for example, is a custom said to have originated in AD 700. Or the equally ancient Pig Bell, which is rung every night at 8 p.m. from St Peter's Church in Sandwich to signal that pigs may now be released into the streets to eat the rubbish.

It is remarkable that the hundreds of different folk customs – some highly publicised, others scarcely known – have defied the swings and roundabouts of history, to survive more or less intact into the twenty-first century. Henry VIII disrupted many of the quaint, semi-religious customs during the Reformation, then Cromwell and the Puritans put a stop to all festivities through the Commonwealth – until these were encouraged back again at the time of the Restoration. The later

Agricultural Revolution saw landlords and farmers discouraging the old seasonal feasts and customs on the grounds that they had the potential to compromise efficiency and waste man-hours. The infrastructure of many rural communities was then destroyed by the eighteenth-century Acts of Enclosure, when displaced farm workers were forced to seek work in the growing industrial towns, and the coming of the railways was believed to herald the death knell of the pastoral way of life.

Villagers freed from their protective isolation would abandon themselves to urban vices, claimed the social moralists. Depraved townspeople hearing of some simple rustic festival, would descend in trainloads and transform it into 'A pandemonium of vice and drunkenness', according to the nineteenth-century British writer James Greenwood. Two world wars were followed by rapid advances in agricultural technology, which were responsible for removing whole generations of village people as machinery replaced manpower. Successive socialist governments sneered at our heritage of customs and traditions, claiming they were archaic relics of the class system.

At one time or another our customs have been adversely affected by political, religious or social influences. Some disappear altogether in one generation, only to be revived in the next. Today, with the revival of interest in our rural heritage, these customs are stronger than ever. The reason for their popularity remains the same as it was a millennium ago; customs and tradition are essential to any community. They engender a feeling of individuality, collective pride and the sense of historic security which continuity has always provided. There is also another important sentiment in this age of high-speed modernity: only when something is in danger of being lost do we appreciate its value.

The Jethart Ba'

One of the more unusual customs takes place in the nearest town to our farm. On the first Thursday after Shrove Tuesday, the shopkeepers and townspeople of Jedburgh can be seen frantically boarding up their windows. Every year, at this small town in the Scottish Borders, celebrated for its Royal associations, exquisite twelfth-century ruined abbey and medieval reputation for rough justice – 'Where in the morn men are hung and drawn, And sit in judgment after' – an extraordinary ball game is played.

The Jethart Ba' consists of two teams – the Uppies, men born above the Mercat Cross and the Doonies, those born below it – competing for possession of a leather handball. There are no rules, other than that the ball must not be kicked, or any sort of defined playing area. The Uppies endeavour to get the ball to their *Hail* on Castle Hill and the Doonies, to theirs at the bottom of town. Anyone can join in and an enormous struggling, cursing, scrum of men surges up and down the streets and alleys of Jedburgh; in and out of gardens; over the A68 into the River Jed and back into town again, leaving a trail of minor injuries and property damage in its wake.

Local legend insists that the origins of the game started in the fifteenth century, when some of the Jedburgh callants, or young men, returned from one of the interminable cross-Border scuffles that were a feature of the period, triumphantly clutching an English soldier's head. This was thrown down in the Market Square for all to see; someone gave it a kick, someone else kicked it back and all of a sudden, a grisly game of football started. Everyone had such fun that the townspeople decided this should become an annual event, with the Englishman's head replaced by a leather ball stuffed with straw.

By 1704, the game had become so violent that the town council forbade 'the tossing and throwing up of the football at Fastern's E'en within the streets of the Burgh' because this had 'many times tended to the great prejudice of the inhabitants … there have been sometymes both old and young near lost their lives thereby.' By now, though, the annual ball game was deeply embedded in the consciousness of Jedburgh's male population and it was unthinkable that it should be stopped. A compromise was reached by substituting a handball, with coloured ribbons attached to represent the Englishman's bloody hair, for the football.

The game continued with unabated enthusiasm until 1849, when a more than usual amount of property damage and personal injury occurred. The Burgh authorities decided that enough was enough and banned the game altogether, imposing ruinous fines on anyone found playing it. They little realised how much the people of Jedburgh enjoyed wrecking their town once a year, or that anyone damaged in the game wore the injuries with immense pride. Jedburgh would not be Jedburgh without its traditional Shrovetide Ba' and so the matter was put before the High Court in Edinburgh. A ruling was secured where the rights to play the game through the streets of the town, were sanctioned by 'Immemorial Usage'.

Fighting Shrews

Behind the old stone-built, slate-roofed steading, with its cattle byres, cart sheds and stabling for the farm horses, is a small, steep hill. This must have been replanted about a hundred years ago with larch and pine, amongst which are one or two of the original oaks and some big ash. The trees were thinned in the 1970s and, too difficult to extract, the timber was left to rot. Other trees have come down in storms over the years and beech, alder, birch and ash have self-seeded in the gaps. The floor of the wood, in among the jumble of rotting trunks, is covered in mosses, ferns, bracken, grass, briars and little anemones. One way or another there are tons of decaying matter, providing habit for every sort of insect and invertebrate. In April, the wood is a riot of small birdsong. Walking through here the other day, I could hear above the endless pinging of siskins, great tits, coal tits and little undulating long-tailed tits, angry rattle of wrens and the hollow tapping of a great spotted woodpecker, a series of shrill, frenzied, squeals.

Edging closer to the noise, I spotted two tiny, long-snouted, common shrews furiously fighting among the leaves at the base of an ash sapling. Except when mating, both sexes are solitary and extremely aggressive. Common shrews have a home territory lasting their lifetime, of around 400 square metres in woodland and half that in pasture, every inch of which they belligerently defend. Should two strangers of either sex chance to meet out of the breeding season, they freeze momentarily in outraged disbelief, before rearing onto their hind legs, shrieking with anger. A brief but ferocious battle ensues with much scratching and snapping of teeth, generally ending with the trespasser being driven off and neither protagonist badly injured. It is a very different matter during the breeding season, April to September, when male shrews roam outside their territories hunting for a mate. Now, the sheer savagery and determination of their testosterone-fuelled encounter, becomes a fight to the death.

Cousins of the minuscule pygmy shrew, Britain's smallest mammal and the beautiful water shrew, common shrews are well distributed all over Britain except Ireland and on high, peaty moorland in the north of Scotland. They live among long tussocky grass, roadside verges, hedge bottoms and in deciduous woodland. Shrews have a high metabolic rate, like moles and expend so much energy in their craving for food, they would starve to death if they did not eat every three hours. Rarely seen, a shrew hunts by day and night, moving in a jerky run, twittering and muttering as it pokes its long nose into crevasses, under logs, in among leaf litter or standing on its hind legs to snatch at a passing insect. Shrews will eat any invertebrate – pouncing on slugs, snails, spiders, beetles, centipedes and scratching into dead wood for woodlice. Such is shrews' need for food, they will, if desperate, eat carrion and other shrews. Their favourite diet is earthworms which they hunt along tiny burrows they dig just below the surface, relying on their acute sense of smell and bristling whiskers, which enables them to travel along tunnels in the dark, to guide them towards prey. Every couple of hours, they collapse through exhaustion and snatch a few moments of deep sleep before being impelled to continue hunting. All British shrews are known as 'red-toothed' from the iron deposited in the enamel of their crowns, which makes the teeth more resistant to wear and tear. Nature has also given them another advantage in the form of a toxin in their saliva which stupefies their prey.

Although only 7.5 cm long, shrews are Britain's most abundant mammal, with over a hundred to the hectare in woodland and about fifty in grassland. The Mammal Society estimates that there are nearly 42 million of them. Almost black in their winter coat, which becomes lighter after the spring moult, the two sexes are only distinguishable by a patch of white hairs on the back of the female's neck where fur has been torn out by males during mating. Female shrews have between two and four litters of up to ten young, between April and September. The kits are born bald, blind and defenceless and she rears them totally without help from the male, in a well-hidden nest of grass and leaves, until they become independent at about a month. During that time, she must find twice the normal amount of food, a task which seems almost impossible. If the nest becomes threatened in any way, the mother shrew will lead her young to safety in a caravan fashion, with each using their mouth to hold on to the tail of the one in front.

There is a very high mortality rate among young shrews, probably due to competition for food at the end of summer when the population expands significantly. For a creature so aggressive, they are susceptible to stress and can die of shock. Country people believed shrews could be killed by thunder. They are vulnerable to a large number of predators, weasels, stoats, foxes, domestic cats and dogs. Although these will kill shrews they rarely eat them, due to the foul-smelling scent gland contained on their thighs, primarily used to mark out their territories. Birds of prey, which have a poor sense of taste, and owls in particular, do eat shrews and in some parts of the country, badgers are so desperate for food, they have now started to scavenge for them.

For centuries, shrews suffered from completely unjustified persecution. Their Latin name, *Araneus* means spider, referring to the old belief that they were poisonous. No doubt this was based on the disproportionately painful bite these little creatures are capable of inflicting and was to lead to dire consequences for shrews. The warmth from cattle and sheep lying down to chew the cud, attract a variety of flies and insects. These in turn attract hungry shrews and hedgehogs. Where hedgehogs were accused of stealing milk, 'poisonous' shrews were the cause of mass outbreaks of lameness, one of the dreaded 'murrains' that periodically devastated livestock. Gilbert White, in *The Natural History of Selborne*, describes the common superstitious cure of the time. A hole was bored into an ash tree and a live shrew forced in. The hole was blocked up and, a twig or branch cut from the 'shrew ash' applied to afflicted limbs.

I was wondering how to separate the panting, scuffling shrews without injuring them before they killed each other, when a young sheepdog I had with me intervened. This year-old collie pup was at the stage of goofy immaturity where harming a fly would be beyond his mental capabilities, but now he crept forward trembling, his nose dribbling with confused excitement. Standing over the shrews, as they writhed and tugged at each other, he slowly lowered his head and inhaled deeply. The pungent smell from their hot scent glands caught him and he reeled back, exhaling violently. It had rather the same effect as chucking a bucket of water over two fighting dogs. Covered in canine mucous the shrews disengaged, exchanged horrified glances and scuttled off in different directions.

Charcoal Burners

Immediately after the Second World War, my father left his regiment and bought a farm in that lovely part of the High Weald in East Sussex, on the northern edge of the Ashdown Forest. This is a landscape of rolling hills, sandstone outcrops, little streams running through steep-sided ravines, scattered farmsteads with small, irregular-shaped medieval fields linked by sunken lanes and paths, amongst areas of ancient broadleaf woodland.

Some of my earliest memories are of my sister and I being taken on afternoon walks through the woods on the farm by our nanny, the redoubtable Nanny Pratt. The woods were a mix of coppiced ash, hornbeam and sweet chestnut, known as underwood, growing in clusters from single stools, and individual oak standards scattered about, trees allowed to grow for their timber without being coppiced. Nanny Pratt always carried a trug on our walks to put wild flowers in for the nursery or edible plants, berries and nuts. In the early spring, she would look for wild garlic plants growing in damp glades or the banks of the little streams, where the water ran reddish-brown from iron ore deposits in the local clay. By May the woodland floor was a carpet of bluebells, wood anemones, wood sorrel, woodruff and shiny leaved dog's mercury, wild arum, white hellebore, little purple orchids and wood spurge. As the summer wore on, herb bennet, primroses, foxgloves, figwort, meadowsweet and purple-flowered enchanter's nightshade could be found.

The woods were a haven for wildlife and a constant source of delight and fascination. A stick poked among leaf litter would be guaranteed to produce something of interest; a disgruntled ground beetle, his glossy blue-black carapace glinting in the sunlight; a longhorn beetle with waving antennae or a thrilling, fast-moving wolf spider. Even woodlice and millipedes had their entertainment value. There were hoverflies, brightly coloured weevils, caterpillars, and where the sun shone through the overhang, any number of beautiful woodland butterflies – white admirals, purple emperors, commas – and a whole range of woodland fritillaries. Nuthatches or tree creepers scuttled up and down the trunks of the old standard trees and we would often hear the rasping curse of a jay or the silly laugh of a green woodpecker. Cock pheasants might be seen scratching for food on one of the bridle paths and there was always a background of twittering, whistling little woodland birds such as chiffchaffs, warblers, tits, robins, wrens and blackcaps. We might see a hedgehog rootling for slugs, an adder curled up asleep on a sunny bank or find ourselves watched by a timid roe deer. In the autumn, when the leaves turned golden and the ferns began to die back, fungi would appear. Clumps of yellow honey fungus, beefsteak, velvet shank or beech tuft on old stumps; oyster mushrooms, chanterelle, boleti, Caesar's mushrooms, morels, puffballs and sometimes fly agaric and death cap. To all of these, Nanny Pratt would cry, 'Don't you go near them.' Autumn was nutting time, when the trug was filled with clusters of hazelnuts in their little green caps, sweet chestnuts or blackberries and rose hips. Winter was my favourite time of the year. I loved the silence of the woods, the long shadows and the stark eeriness of bare trees, the red, citron, russet, black, bronze or copper of fallen leaves and the musty, mouldy smell of decay.

Sometimes, when the frost was hard on the ground, we would smell woodsmoke and come across the farm men coppicing chestnut trees for fencing posts and burning the trash. Chestnut poles would be cut to length and stacked in 'cords' to dry until the following winter, or dry poles from the previous year split with sledgehammers and chisels, and then loaded onto a four-wheeled box wagon, whilst the carthorses stood patiently in the shafts. The woods were coppiced on a rotational cycle and the areas to be cut were known as *coups*. Hazel was coppiced every six to eight years, chestnut every ten to fifteen years, ash every twenty, hornbeam twenty-five, and oak, around fifty years.

One winter, a family of charcoal burners set up camp in Drew's Rough, a hornbeam wood to the north of the farm, cutting and stacking the wood to dry for burning. Their arrival was an endless source of rumour and gossip among the farm men; charcoal burners lived in huts made of turfs and were worse than gypsies for thievery and poaching. They ate badgers, hedgehogs, squirrels, snails and little woodland birds which they caught with birdlime made from fermented holly bark; were immune to the bites of adders, which they caught with their hands, skinning them and selling the fat to people who believed it cured deafness and rheumatism. I thought they sounded fascinating, and my one ambition was to be taken to visit the camp, but needless to say, as far as Nanny Pratt and indeed everyone else on the farm was concerned, Drew's Rough was a place to be avoided.

In the early spring, Nanny Pratt took her annual fortnight's holiday and was replaced by a Miss Knowles, a young woman from an agency. By then, all anxieties over the charcoal burners had been forgotten. They kept to the wood and were rarely seen; nothing had been reported stolen, nor had any of them been caught poaching. With my sister away, there was no one to contradict me when I suggested to Miss Knowles that Drew's Rough would be a pleasant place for our afternoon walk. I did not actually know where the charcoal burners were working, and I suspect that my curiosity would have been satisfied by simply watching them unobserved from the safety of the trees, if we ever found their camp.

As it happens, one moment we were ambling along the old, sunken, leaf-filled drove road that ran through Drew's Rough and the next, we came round a bend to find ourselves in a clearing, where the filthiest man I had ever seen was shovelling earth onto a circular pile of logs. A horse and cart laden with cut cords was being led down a track into the clearing by another man, whilst a third loaded cords from a stack onto a wheelbarrow with four uprights rather than a body, known as a 'charcoal burners' mare'. There were the remains of a fire that had burnt down, with a heap of hessian gunny sacks filled with charcoal beside it, partially covered by a tarpaulin. Across the clearing, where the ground rose, were several huts made of poles, turfs and canvas, shaped like wigwams. A blackened cooking pot hung on a tripod in front of them and nearby, two grubby children played with a scruffy mongrel chained to a stake.

Until now, the men had been concentrating on their work and we might have slipped away unnoticed, but the dog saw us and started to bark; two women emerged from the huts and everyone stopped and stared at us. We couldn't leave now, so we walked across the clearing to the man with the shovel and the two women (Joe Botting, our gardener, told me later that charcoal burners' women were known as 'motts') came down to join him. He wore the greasy wreckage of a trilby hat, Derby tweed trousers, hobnailed boots, a collarless flannel shirt and an old waistcoat. Everything – his unshaven face, neck, arms and his clothes were black with charcoal grime. The two women were no cleaner; one was lank and sinewy, with dirty black hair hanging in a rope down her back, wearing a threadbare jersey, a man's jacket and a thick tweed skirt, with her skinny bare legs thrust into a pair of cut-off gumboots. The other wore oily dungarees and an army greatcoat. Both of them, like the man, had dark obsidian eyes, smelt of stale woodsmoke and seemed to have soot ingrained in their skin.

'Good afternoon,' said Miss Knowles. 'We're from the farmhouse. This is Major Scott's son,' succinctly establishing our credentials.

This was met with a stony silence, whilst the three of them stared at us unnervingly. Then the skinny woman said in a high, rasping voice, 'Ooh, look at 'im. Look at his hair, innit booful?' In those days, I had a mop of blond curls which Nanny Pratt made me wear rather on the long side. I loathed them, because it made me look girlish. The skinny woman seemed to agree; 'Innit a crime for a boy to have hair like that? What I wouldn't give for curls like them.' At this, she gave what was no doubt intended to be an encouraging smile, exposing the stumps of her teeth and stretching her hands with their long, blackened fingernails towards me, attempted to fondle my hair. At that, my nerve broke and I ran squealing out of the clearing with her horrid cackling laughter ringing in my ears, back down the old drove road towards home. For years afterwards, I was plagued with nightmares of those hands reaching out for me and the clumping footsteps that followed as I ran through the leaves.

Pigeon Racing

We are a nation of sportsmen, more sporting, I like to think, than any other and there is nothing we love more than a challenge, particularly where animals are concerned; racehorses, greyhounds, trail hounds, trotting ponies, terriers, ferrets or, the working man's favourite, racing pigeons. 'A flat cap, greyhound shit and pigeon dung,' was a music-hall reference to the principal components in the life of an urban sportsman, depicted so succinctly in the cartoon character, Andy Capp.

Pigeon racing was a natural progression from the centuries-old practice of using domesticated rock pigeons, which have an innate ability to find their way home over long distances, to carry messages. Until the development of the telegraph system, carrier pigeons were the swiftest method of communication and their use continued throughout both world wars, when over 200,000 carrier pigeons were in use, thirty-two of which were subsequently awarded the Dickin Medal for bravery, the animal equivalent of the VC. Pigeon racing as a sport began in Belgium in the early nineteenth century, with the first organised race, held over 160 km, in 1818. This was followed in 1820 by a race of almost double the distance, from Paris to Liège, and in 1823, from London to Antwerp. In Britain at that time the sport was largely confined to private matches between the sporting gentry and endurance races, with the record for the longest journey ever flown still held by a pigeon belonging to the Duke of Wellington. This bird was released from a sailing ship on 8 April 1845, near Ichaboe Island off the west coast of Africa, dropping dead fifty-five days later on 1 June, only a mile from its loft at Nine Elms, Wandsworth, having flown a distance of 18,065 km, including a 5,700-km detour over the Sahara Desert.

Keeping pigeons soon became a popular urban backyard hobby that could involve the whole family; this and the expansion of the rail network in the mid-nineteenth century meant that the sport of racing took off in Britain, as it was now possible to transport pigeons to the start point of a race practically anywhere in Britain or on the Continent. The sport received another upsurge in popularity in 1886 when King Leopold II of Belgium gave a pair of racing pigeons to Queen Victoria, who established a racing loft at Sandringham. Both Edward VII and George V were keen fanciers and had success with the Royal pigeons, including first prizes in the national race from Lerwick in the Shetland Isles. Pigeons from the Royal Loft were used as carrier pigeons during the First and Second World Wars; one of which, a bird named Royal Blue, won a Dickin Medal for its role in carrying a message reporting a lost aircraft in 1940. After the war, the pigeons returned to racing, winning further national and international races.

The late Queen Elizabeth II took an active interest in the Royal pigeons and regularly visited the lofts – new ones were built in 2015 at a reputed cost of £40,000 – when she was at Sandringham, where 160 mature birds and 80 juveniles were managed and trained by the official Loft Manager for Her Royal Majesty the Queen. The manager entered pigeons into one or two club races each week and all national races during the season, which runs from April to September. At various times over the years, the Queen's racing pigeons, with their distinctive leg rings bearing the Royal insignia, won every major race in the calendar. In July 2008, Her Majesty donated three birds from the Royal Loft to the Wetherby Young Offenders Institute, to give prisoners the opportunity of learning a constructive hobby in the hope that some might continue it on release and reduce the likelihood of re-offending.

As with all country sports, pigeon racing has an eclectic following and the six regulatory organisations in Britain have memberships that range from peers to the sport's traditional urban heartland members, and everyone in between.

Young birds are trained by taking them further and further away from their lofts until they are ready to compete in novice races of 100 km, which increase to 1,000 as they mature, with big international races covering distances of 1,600 km. A race is won by the bird that flies the fastest number of metres per minute, with the exact distance from the race point to the competing member's loft calculated by computer. Birds will reach speeds of 130 km an hour at stages over a race if the wind is behind them, and about 80 km an hour on a calm day.

The size of the average pigeon loft is between 50 and 100 birds, although some fanciers, such as the Masserella family, who became market leaders in breeding and selling racing pigeons, will have several thousand birds, some of which will be extremely valuable. A racing pigeon can sell for anything from £50 to £150,000, depending on bloodlines, and in the Far East, where millions of dollars are bet on the outcome of pigeon races, immense sums exchange hands over successful birds. The current record price for a single bird was in August 2019, when two competing Chinese buyers pushed the price of a five-year-old Belgian cock bird up to £1.07 million.

There are around 50,000 people in Britain devoted to pigeon racing, with one-loft racing gaining popularity. This is the process of training birds bred by many different breeders in the same loft, under the same trainer and in the same conditions – as opposed to trainer against trainer in their own lofts and usually with their own birds. It is thought to be the fairest method of proving which bloodline or breeder is best and usually provides the highest amount of prize money. Pigeons are recorded by electronic timing systems scanning the birds as they enter the home loft, with winners decided by as little as one-hundredth of a second. The birds are all taken to the same release point and they return to the same home loft, so therefore it is the fastest bird to complete the journey from A to B. One-loft racing is now becoming very popular all around the world with fanciers able to compare their bloodlines on an equal basis against the many other pigeons.

Pigeon racing is not without heartbreak; there are fatalities caused by vagaries of the weather, electricity pylons and, increasingly, birds of prey – it is estimated that 70,000 pigeons are killed whilst racing every year by raptors.

Lockdown

How incredibly fortuitous that the Covid-19 lockdown was imposed at the start of an utterly glorious spring, with nature at her most benign for decades, giving us day after day of practically uninterrupted warm, sunny weather. The dawn chorus was truly exuberant and wild flower growth quite breathtaking, with verges on country lanes gleaming white as wild parsley, stitchwort, white dead nettle and hawthorn came into blossom. Gorse, buttercups, meadow vetchling, cowslips, celandine, dandelions, yellow rattle and bird's-foot trefoil threw a golden sheen across heaths, wood margins and meadows. Woodlands carpeted in bluebells, clusters of purple-flowered bittersweet, pink purslane, blue forget-me-nots, red campion, wood sorrel, blue speedwell, wild garlic and sweet woodruff, in their last flush of growth before the leaf canopy closes over them.

A haze of yellow, pink or white blossom surrounded laburnum, ash, horse chestnut, crab apple, rowan, elder, sycamore or whitebeams, while in old permanent pasture and damp hollows, ragged robin, poppy, sundew, horsetail and pennywort added a splash of purple, scarlet and brilliant green. Sometimes, when the temperature dropped on a May evening, a warm gust of air passed by, filled with an exquisite scent. A combination of all the day's blossom fragrances, which only lasted for a second or two, then the evening chill descends and one of the wonders of nature is completely gone. It hardly bears thinking how much more difficult life would have been for everyone, if winter had dragged on and spring had been cold, wet and blustery.

Insect life explodes when the temperature rises; millions of every conceivable type of bug, beetle, ant, spider, earwig, bee, midge, gnat, wasp, woodlouse, slug and snail, crawl, squirm, hop or fly, in a frenzy of activity. May bugs thump against windows on warm evenings or blunder into the room, drawn to the light in search for a mate; yellow brimstone, peacock, orange-tip and speckled woodland butterflies are joined by Adonis, chalkhills and common blues; bright-coloured dragonflies defend territories on the edges of ponds and lakes; damselflies bind together as they mate, and in clean, fast-flowing streams, where white flowered, water-crowfoot grows, brown trout rise to the mayfly hatch. Insect activity brings the summer migrants; little blackcaps, white throats, wheatears, willow warblers and cuckoos, the heralds of warm weather. Endlessly twittering swallows, swifts and house martins swoop around treetops feeding on midges carried up by high pressure thermals, whilst sand martins skim back and forth above water.

Wild flowers reach their peak of bloom in the oppressive, airless days of July, when hedgerows, riverbanks and meadows are a riot of colour before they begin to die back. Pink dog roses, yellow honeysuckle and bramble flowers twine through the body of hedges to compete with greater bindweed and old man's beard with their profusion of tiny white flowers. Verges are full of magenta willowherb, pink teasels and feathery-leafed yarrow, blue meadow crane's-bill, yellow rattle, purple butterwort, pink bistort, mauve foxgloves, daisy-like scentless mayweed and scarlet poppies. Old pastures burst with colour; fluffy yellow lady's bedstraw, orange bird's-foot trefoil, ox-eye daisies, purple betony and greater burnet. Pink betony and knapweed, blue devil's-bit scabious, lilac spotted and southern marsh orchids, white mouse ear and the bumblebees' favourite – white and pink clover. Meadow vetchling, water forget-me-not, watermint, gypsywort, brooklime and yellow-flowered flag irises thrive on the edges of streams and rivers. The flower which is it at its finest in July is meadowsweet, which grows in profusion in damp meadows, boggy stream banks, pond and ditch sides throughout Britain. At this time of year, the heavy sweet scent from the foaming clusters of tightly packed, creamy-white flowers is quite magical. We are blessed to have such a variety of beauty on our doorsteps as soon as we leave our homes.

The NHS doctors, nurses, care home staff, ambulance drivers, paramedics and all those in the front line of the heroic fight against the pandemic are rightly applauded for their selfless fortitude and dedication, as are the postmen and all other essential public service employees, who kept us going during lockdown. There is another group of people equally deserving of the nation's gratitude; the custodians of the countryside. The landowners, arable and livestock farmers, woodsmen and foresters, hedgelayers, drystone dykers, river keepers and gamekeepers. Quietly working away as they have for centuries to create and maintain our green and pleasant land, which offers a solace, well-being and sense of freedom to everyone when they escape the confines of urban lives. Britain has a landscape unique in its diversity, from heather moorland to coastal marshes, home to a vast array of different wildlife. Despite the bigotry and ignorance of animal rights activists, we have generations of gamekeepers to thank for their vigilance in protecting the vulnerable from being swamped in a sea of adaptive predators – corvids, foxes, stoats, weasels and grey squirrels.

Two-thirds of all rural land in the UK is keepered and shooting estates spend over £250 million annually on conservation, which provides habitat, food, shelter, roosting and nesting sites to a multitude of little farm and woodland songbirds, whose billing and cooing and fond pursuing is the pantomime of spring. The ground nesters of heath and marshes, and the summer visitors who flock to nest and rear their young safe from predators, in a haven provided by the 4 million acres of UK grouse moors. The golden plover, curlews, snipe, lapwings, skylarks, meadow pippits, merlins, dunlins, redshanks and dotterels, breaking the long silence of winter with their melody. How sad and sterile our countryside would be, if all of this were in any way diminished.

Bracken

Up in this part of the world, April can be a frustrating month. All the indications that spring is nearly upon us are here, without the warmth. Hill lambs, Cheviot or blackface, sheltering behind clumps of reeds whilst the ewes graze. Oystercatchers and dippers, flying in pairs up and down the river. Cock snipe producing their extraordinary throbbing wind music over areas of bog and reed. Green plovers and curlews mewing up on the moor or hanging like crescent moons, making that lovely soporific, drawn-out burble. Alder trees and willows thick with catkins, acres of acrid-smelling wild garlic with delicate, pointed white flowers and among the great swathes of brown bracken litter, insipid tightly curled heads of new growth are beginning to appear.

Where any other green growth is welcomed at this time of year, the appearance of these innocent-looking crosiers is viewed with a jaundiced eye. In June they will be waist-high and in some places, a virtually impenetrable barrier 2 metres tall by August. Bracken is an aggressively invasive weed that runs rampant throughout most of the world except Antarctica and desert regions, covering nearly 11,000 sq km of England and Wales, and half as much again in Scotland, most of it in hill stock rearing and heather moorland. A small area of bracken is a bonus to any sporting landlord as fox and woodcock cover, but once out of control, it is immensely detrimental to man and beast. Reducing potential grazing areas, invading ground-nesting bird habitat and blanketing out competing vegetation, particularly heather. In hot weather, it makes gathering hill stock a nightmare and sheep seek shade inside its dank interior, becoming vulnerable to maggot fly strike. It is poisonous to livestock. Various species of disease-carrying tick hibernate and breed in the mulch at the base of plants and the spores, released when the fronds open in July, cause blindness in animals and are recognised to be carcinogenic. Forestry Commission workers are now required to wear face masks when working near bracken and the Scout Association advises members to 'refrain from walking through bracken patches and from using bracken to construct backswood shelters'.

There are a variety of factors that have contributed to the spread of bracken over the last 150 years. Damage caused by overgrazing and the loss of cover through heather beetle, has allowed bracken to become dominant in many areas. The rhizomes and wind-borne spores travel considerable distances, quickly becoming established on bare ground, particularly if heather is burnt near bracken beds, but one of the main reasons has been losing its value as a natural resource. Bracken once had a myriad of applications and was harvested annually for thatching, deep litter bedding for livestock and as a base for straw and haystacks. Root crops were covered in bracken to protect them from frost, either in the ground or in clamps. A tanning solution called 'brake water' was made from the rhizomes to create a yellow dye for colouring wool and leather – particularly kid and chamois. As a quick burning, rapid heat fuel, it was used in bread and brick making as well as lime burning. During the eighteenth and nineteenth centuries, an enormous acreage was cut to be burnt for the alkali and potash used in glass and soap making – potassium-rich bracken ash was much sought after as a fertiliser for root crops. Garden produce was displayed and kept fresh on ferns in grocers' shops and market stalls and before the invention of pneumatic tyres, anything breakable such as slates, pottery, earthenware troughs and sinks was always packed in bracken for transport.

Bracken has a variety of certain medicinal properties once extensively used in herbal remedies – a decoction made from the rhizomes was recommended by Culpeper to remove stomach worms and for easing bronchitis. Early colonists to America, New Zealand and Australia were slightly non-plussed to find the natives using bracken for the same purpose. A powder made from dried, powdered stalks and roots was used as a coagulant in wounds and dusted on cuts and ulcers. Fronds were cooked up in grease for severe bruising or made into poultices for boils. Sap squeezed from the stalks relieved stings, the inflammation left by tick bites, mouth ulcers and toothache. Smoke inhaled from burning bracken was recommended to relieve headaches and repel mosquitoes. Gardeners watered plants with a solution made from boiled fronds to kill aphids and the hopeful rubbed it into their scalp to promote hair growth.

In certain parts of the world, bracken rhizomes and the emergent fronds, known as 'fiddleheads', have been eaten for centuries. Rhizomes are high in starch and were made into a form of flour by aboriginal people across North and South America and from the Canary Isles to the Far East. Confectionery made from rhizome flour is a speciality of the ancient Japanese city of Nara and the fiddleheads, eaten rather like asparagus or as a purée, are very popular throughout the country. Pickled or fried fiddleheads are traditionally eaten in the Pelion region of Greece to accompany the local spirit, tsipouro and have become something of a cult food in America. Despite concerns about their uncooked carcinogenic properties, fiddlehead greens can be bought frozen or canned from gourmet and Asian food stores. They have never really caught on here – Dorothy Hartley, in *Food in England*, remarks that they taste rather like Darjeeling tea and are either liked very much or not at all. In Russia and northern Scandinavia, mature fronds are added to malt to make a type of beer.

As bracken lost its commercial value, farmers resorted to a variety of control practices, some of which are now employed by organic producers. Repeated cutting early in the season, if you are lucky enough to have level, stone-free ground, weakens the plant as does harrowing, disking or bruising with a hexagonal roller. Burning dead bracken through the winter removes the frost-protective layer of mulch and deep ploughing exposes and degrades the root system. Pigs, folded onto bracken beds through the winter will grub up roots and have much the same effect. Chicken dung, which is very acidic, spread on bracken will kill it, but all were time-consuming at a period when manpower was leaving the land. Bracken soon outstripped efforts to contain it and spread into increasingly inaccessible areas. The majority of large acreages of bracken eradicated in the last fifty years, have been sprayed with the herbicides Asulox and Asulan, either with a boom mounted all-terrain vehicle, or by helicopter on ground that was too steep for tracked vehicles. Even with grant-supported Countryside Rural Stewardship schemes, this is extremely expensive at a time of sliding agricultural incomes and only effective, if followed up with liming or a programme of spot spraying re-growth, with a knapsack sprayer.

With bracken spreading at a rate of 3 per cent a year, serious consideration has been given to some of the qualities that were once recognised by our ancestors. Bracken contains properties that repel insects and inhibit weed germination and the mulch is now sold commercially in garden centres as a winter plant cover. Fronds harvested with modern machinery are a lower cost, more durable alternative to straw bedding for livestock, which degrades into high-quality compost. Bracken is already harvested in North America and Brazil for pharmaceutical purposes connected with the treatment of bronchitis and has a capacity for control of internal parasites in animals. Bracken's biggest potential is as a biofuel with a valuable ash by-product. Research by Aberdeen

University has discovered that the high heat generated by incinerated bracken produces as much energy as coal. Fifty tonnes of dried bracken harvested in June is reputed to produce one tonne of valuable potash and whilst a considerable acreage would be inaccessible, it is estimated that 2.5 million tonnes could be harvested in Scotland alone. With so much available, it seems a shame not to make use of it.

Summer

Soft or Hard Grouse?

The uplands of northern England and Scotland have their brief moment of glory in August, as Britain's 11 million acres of heather moorland – 75 per cent of the world's remaining resource – burst into honey-scented flower. There is no finer or more magnificent sight than the great swathes of purple-clad hillsides, broken here and there by the black mosaic patterns of this spring's managed burning, where in the words of Robert Burns, 'the moorcock springs on whirring wings'. An extraordinary little creature, the red grouse; endemic to Britain, the finest-driven game bird in the world, hardy, unpredictable and blissfully unaware of the astonishing amount of attention it attracts. No other animal on earth can receive the same degree of annual press coverage – both positive and negative – or have such a volume of learned literature devoted to it. None can cause such swings of elation or despair, ecstasy and anguish; have such anxiety expended over its well-being; acquire the mystique and social status that draws sportsmen from all over the world; be responsible for a landscape rich in colour, wildlife and biodiversity, or provide a vital revenue to some of the poorest and under populated areas in the country. If red grouse were to disappear completely, the economic impact would be incalculable.

Hardly surprising, that such a remarkable bird should be the source of considerable differences of opinion as to its relative merits north or south of the Scottish Border, to the extent that there is even a body of opinion who insist that one Scots grouse is worth four Sassenachs. This is far too broad a generalisation to be taken seriously, but nevertheless there is an assumption among some people that Scottish birds are harder to shoot than English. When one takes into account the distance and variation in climate between a grouse moor in Inverness and one 400 miles further south in Derbyshire, this hypothesis seems to justify further scrutiny.

I asked the opinion of Robert Rattray, the senior partner of Galbraith Sporting Lets in Perth, with his extensive knowledge of grouse moors throughout Scotland, whether Scottish birds were a hardier breed than those south of the Border. 'It all depends where you are of course,' he told me. 'Perhaps on the more extreme areas of the North East, where the ground is higher and steeper, the grouse could be said to be tougher and fly harder, simply because everything is that much harsher: the terrain, the peaks and troughs of weather, severity of the winter, shorter growing season and a higher mortality rate, leads to the survival of the fittest.' These are the heart stoppers that come corkscrewing round the side of a 3,000-foot hill when one is in a butt halfway up a near vertical slope on somewhere like Invercauld, Edinglassie, Candacraig or Dorback near Tomintoul.

Does weather and location make Scottish grouse a harder flying bird compared to English? Jonathan Kennedy of CKD Property Advisers and one of the finest grouse shots of the age, tells me that with a couple of exceptions, grouse on moors in the North Pennines, such as Gunnerside, Holwick, Wemmergill, or Weardale, are as good as any in Scotland. Having seen hordes of grouse with a breeze behind them coming towards the butts at High Crags on Wemmergill, 2,400 feet above sea level, I would have to agree.

Topography is everything; grouse are territorial and hefted to certain areas within a moor, and how they adapt to flying in their natural environment – steep or flat ground, will influence the way they perform on a driven day. Loosely speaking, grouse are 'softer' on lower altitudes, but there are moors on both sides of the border where the grouse are 'tame' and indeed, there are drives or lines of butts within moors, where grouse are presented to provide completely different shooting challenges. Mayshiel in the Lammermuirs is an example; the ground is undulating and gently sloping where Mayshiel itself marches with the Earl of Haddington's Johnscleugh and grouse here seem to skim inches above the heather at unnerving speed, leaping 20 feet into the air as they reach the butt line. Move through the estate to the beats on Fasney and you are in an entirely different landscape, which could be anywhere on any of the steepest ground in the Highlands. In particular, the famous line of almost perpendicular butts known as Cardiac Climb, where the grouse come at all angles, swooping like swallows round the side of the hill and whipping over guns at the last moment.

Simon Thorp, the director of the Heather Trust from 2002 until 2018 and now a consultant on various moorland projects, agrees that grouse are perhaps a little hardier in the North East – the temperature on the high ground in Aberdeen on 6 May this year when hens were nesting, was minus 5 degrees Celsius compared to a balmy 12 degrees in the Peak District. Cold weather at that time of year is not going to bother a hen grouse, provided it warms up in time for insect activity at hatching, but southern grouse might hatch a few days sooner and have a slight edge over northern ones with earlier plant growth and increased nutritional value. To say Scottish grouse are harder to shoot is too sweeping a statement, because there are such a variety of regional and local behavioural differences.

Grouse on moors such as Danby, Egton, Rosedale, Spaunton, Bransdale, Commondale, or Snilesworth, on the undulating plateau of the North York Moors are generally considered tamer than elsewhere and rarely pack as the season progresses. I asked George Thompson, the head keeper on Spaunton for twenty-eight years, the reason. 'It is because they see so many people,' he told me. 'Spaunton is 7,000 acres and we have over 34 miles of busy public footpaths,' but terrain has an influence. The Derbyshire Peak District has just as many members of the public every day of the week and yet the grouse on the western side of the ridge running through the Peak District are known to be sharper on their toes early in the season than the eastern side. Richard May, a board member of the Heather Trust and the shooting tenant of Peak Naze, Mark Osborne of J. M. Osborne, Jim Sutton, the head keeper of Woodhead and Snailsden, and Richard Bailey, the head keeper of Goyt and Crag, all agree that the difference is entirely down to topography.

The western side is altogether wilder and much more rugged, with narrow contours, steep valleys and ghylls, and butts strategically placed to make the best use of them. Moors on the eastern side where it drops down to the River Don, such as the Duke of Rutland's Moscar, tend to be flatter and grouse perform differently. Not that this means grouse being brought in off lower, undulating ground are necessarily any easier to shoot – you may be able to see the beating line in the distance on a flat moor and black dots of grouse, sinking and rising as they hug the contours towards you, but they are just as hard to hit when they jump out of dead ground at 60 mph, as they are when they come skittering and spiralling over the side of a steep slope with a short skyline.

For many years there has been a suggestion that genetic differences exist between Scottish and English grouse, which has an influence on the way they behave. I asked Dr Adam Smith, the GWCT Director of Policy in Scotland, whether there was any scientific data to support this, and he told me that he had found no reference to tests between English and Scottish grouse for genetic differences that would indicate they were a subspecies. Tests between Ireland and Scotland, and Scotland and Norway had been carried out which established various genetic differences between the grouse and while none actually say subspecies, they are strong enough to be close.

If these differences exist, then it is likely similar differences will be present between England and Scotland, but whether it is enough to establish a subspecies is a question that has been argued inconclusively for decades in numerous scientific papers. 'There are clearly physical characteristics that vary between the red grouse populations,' he told me, 'the almost coal-black cock birds of Donside and Deeside, as opposed to the paler birds of the Western Isles; the often very pale winter feathering on the bellies of Scottish grouse rarely seen on English ones. But are these subspecies? Probably not.' Being hatched a few days earlier can make English grouse seem slightly larger and heavier than some Scottish birds – the effect of better-quality food, as English heather is generally 30 per cent more nutritious than Scottish – but that is by no means always the case, as a variety of management and weather factors would have an influence from moor to moor.

On balance, there does not seem to be a definitive answer, but Simon Thorp made an interesting observation, that Scottish grouse are perceived to be harder to shoot than English, because of the romanticism and magic which has been such a major part of field sports in Scotland for virtually the last 200 years. The whole adventure of going up to the Highlands: coming up from the south, or from America, Europe, Russia or Asia, arriving in northern Scotland to shoot in such a unique, stunning and majestic landscape. The spirit of this was brilliantly portrayed in two magnificent oil paintings by George Earl (1824–1908). Commissioned in 1893, *Going North* captures all the bustle, excitement and sense of imminent departure as a party of sportsmen, with their wives, servants and gun dogs wait on Kings Cross Station beside piles of leather trunks, gun cases and rods, to catch the train north at the start of the season. In *Coming South,* the same party wait to board the train at Perth Station a month later for their return journey and now among the luggage are heads, salmon, blackcock and grouse. Funnily enough, there is a similar analogy between the stalking in say, Rosshire and the equally challenging stalking in Cumbria, a mere three hours from London by train.

Perhaps it is all in the mind, but there is one thing all grouse shots will agree on: whether the moor be north or south, high, low, steep or flat; when grouse are coming towards you, he who hesitates is lost.

Slippery Eels

The extraordinary lifecycle of our freshwater eels, about which there is still some uncertainty among marine scientists, has always fascinated me. Mature eels of about fourteen years of age gradually begin to change colour from olive-brown to silvery-grey in the autumn and, during the winter months, vacate their freshwater habitat to head for the sea, travelling overland if necessary. At this stage they are at least 50 cm long and very fat. When they reach saltwater, their gut dissolves and body fat alone must sustain them for the 6,000-nautical-mile ocean crossing to their breeding grounds, believed to be in the Sargasso Sea. Here in the vast, floating reed beds south of Bermuda, eels that have survived the journey mate, spawn and die. Over the course of the following one to two years, billions of tiny, 5-cm leaf-shaped, translucent, larval eels are carried by currents across the Atlantic, arriving at European rivers such as the Gironne, Severn or the River Bann in Northern Ireland, between March and May. By now they will have grown another couple of centimetres, are the shape of a pencil and, because of their translucence, are known as glass eels or elvers. These migrate inland to find freshwater habitats and once their destination has been reached, the elvers become golden-brown and the cycle begins again.

Bait fishing for eels is enormous fun, particularly by the light of a full moon in the autumn and they are some of the best eating of any fish. Because of their ease of drawing to a bait, eels were probably the first freshwater fish caught by Neolithic people, trapped in wicker fish traps placed at the edge of reed beds on rivers, lakes and ponds, or speared with primitive wooden leinsters – eel spears. The demand for fresh eels fuelled a considerable niche market cottage industry at one time, which escalated when jellied eels became popular fairground and racetrack fare. Eels have rather fallen from favour here, whilst continuing to be enormously popular in Europe, with the main supplier of eels to the Continent being the Lough Neagh Fishermen's Co-operative Society, of Toome Bridge in Northern Ireland, one of the last remaining commercial wild eel fisheries in Europe. For many centuries, generation after generation of the same families, mainly from the scattered rural parish of Duneane, traditionally fished the Lough for brown eels with long lines during a season which runs from 1 May to 8 January. The silver eel weirs at the mouth of the Bann operate within a season running from 1 June to the end of February.

Eels become active and start searching for food – carrion, or anything smaller and slower than themselves at night. In the early afternoon, long-liners, working in pairs from deep-keeled open boats, start setting the three long lines, each with 500 hooks laboriously baited with mealworms, earthworms or pollen fry. These are then attached to buoys until four o'clock the following morning, when the fishermen are allowed, under the strict eel-fishing rules on the Lough, 'to take the catch'. The eels are deposited live in cages at landing quays round the Lough for collection later in the day, whilst the fishermen begin the painstaking process of baiting 1,500 hooks and coiling the long lines on setting boards ready for the afternoon.

It is a hard, idyllic life of long hours and periods of unremitting labour, with a history of gradually escalating conflict between the 350 fishing families dependent on brown eels for a subsistence living and the owners of the silver eel weirs at the mouth of the Lough. In 1605, James I granted the ownership of the bed, water and fishing rights contained within the 400 sq km of Lough Neagh to Sir Arthur Chichester, the Lord Deputy of Ireland. These ancestral rights passed to his descendants, the

Earls of Donegal, and subsequently by marriage in 1857, to the Earls of Shaftesbury, the current owners. The commercial fishing rights to the different species were let to interested parties and as long as supply exceeded demand, there was little cause for hostility between the weir owners who leased the eel fishing and the fishing families who believed they had a moral right to continue their historic way of life. The trouble started when the value of silver eels rocketed in the mid-nineteenth century, when those with the fishing rights felt the activities of the long-liners was a threat to the potential silver eel catch at the weirs.

For decades, efforts were made to restrict the eel fishermen, either through litigation or violence, until 1963 when the High Court of Northern Ireland confirmed the absolute right to the eel fishing on the then leaseholders, a consortium of Dutch and Billingsgate fish wholesalers. The fishermen could now only operate with the leaseholders' permission and subject to whatever restrictions they chose to impose. At this low point in the lives of the eel fishermen, Father Oliver Kennedy stepped into the ring. As curate of Duneane, where many of the fishing families lived, Father Kennedy was very aware and deeply concerned at the hardship visited upon his already disadvantaged parishioners. With his guidance and advice, the fishermen formed themselves into a cohesive body that bought the fifth share of the leaseholders' consortium in 1965, when one of the wholesalers put his share on the market. By 1971, with Father Kennedy as managing director, the Eel Fishermen's Co-operative had acquired, in a series of astonishing corporate moves, the entire eel-fishing rights on Lough Neagh and established lucrative new markets on the Continent. In under ten years they had achieved the forlorn hopes of previous generations.

From the outset, the object of the Society has been to safeguard the local community and preserve eel stocks for the future, and here Father Kennedy was ahead of anyone else. Well aware that European wild eel fisheries were overfishing to cash in on demand, and sensitive to the first indication of a drop in returning elver numbers, Father Kennedy introduced fishing restrictions and began buying elvers from the Severn Estuary to maintain stocking levels. Now, in 2009, when a shift in the course of Gulf Stream or the effect of global warming has seriously affected the numbers of elvers returning to European waters, Lough Neagh has the only productive and sustainable wild eel fishery in Europe.

Golden Gorse

The uplands have their brief moment of glory in August, as Britain's 500,000 hectares of heather – 75 per cent of the world's remaining resource – burst into honey-scented flower. There is no better sight than the great swathes of purple-clad hillsides, broken here and there by the black mosaic patterns of this spring's managed burning and periodically, a splash of brilliant yellow where patches of western gorse grow, creating a glorious purple, gold and black carpet, sprawling across the landscape.

There are three species of gorse in Britain: *Ulex gallii*, or western gorse, which grows to about 3 feet high and is found on upland heaths and moorland; *Ulex minor* or dwarf gorse, growing predominantly on the downs and cliffs of our southern coastline and *Ulex europaeus*, or common gorse, the most prevalent and invasive, which was introduced, probably by the Normans, as a source of winter feed for farm animals. The practice of cultivating and harvesting gorse was widespread across Britain well into the nineteenth century and in remote places, for much longer. With a protein content equivalent to half that of oats – about 12 per cent – 2.5 hectares of gorse would provide winter feed for four cows or six horses, whilst dairy cows fed on gorse were said to produce the sweetest milk.

Seeds broadcast on sandy or light, free-draining soil were ready for grazing by sheep in two years or cropping by coppicing in four. Branches were cut with long-handled loppers and taken back for crushing to remove the sharp spines. On some farms, this was done with a broad-headed mallet, flat on one end and serrated on the opposite. Others had purpose-built horse, oxen or water driven whin mills and a good example of a water mill can be seen at the National History Museum of Wales at St Fagans, Cardiff. Many used a stone roller dragged back and forth by carthorses over gorse lain on a bed of thick flat stones and not far from here an old gorse roller is used as a gatepost. For cattle and sheep, gorse needed to be ground down to a moss-like consistency and fed before it fermented. The process for horses was less time-consuming and only involved breaking the gorse up through a mangle or sturdy chaff cutter. I have sometimes cut a bundle of gorse and hung it in a horse's loose box. They obviously find it very palatable and despite the sharp spines, will happily pick away at in preference to hay until there is nothing left but the bare branches, stripped of bark.

European gorse was considered so valuable that early colonists took seeds to fifteen different countries – particularly America, Australia and New Zealand – but it had a whole variety of other uses other than animal feed. If regularly cut, gorse makes good hedging for holding stock and is a useful method of keeping deer and hares away from young trees. The oily wood has a high heat content and, in some areas, was deliberately left to grow to its maximum height of 10 feet to be cut as fuel for bakeries, brick and limekilns. The alkali-rich ashes were used as fertiliser, as a lye for washing cloth or in soap making. Flowers were picked to make a beautiful yellow dye and were said to be good for acne when mixed with goose fat and made into a face cream. Their rich fragrance makes an unusually delicious, sweet wine and for years, the Benedictine monks on Caldey Island, off the coast of South Wales, have manufactured a famous scent from the blossom. The effluent of gorse steeped in water was believed to have insecticidal properties which killed household fleas.

Cultivating and harvesting gorse was time-consuming and labour-intensive. As manpower was replaced by machinery, the management of gorse ceased, and gorse spread over considerable tracts of land through natural, unchecked regeneration. Every summer when the ripe pods explode, seeds are propelled a distance of 2 to 3 yards to form immature seedlings the following year. Although European gorse is sometimes used in land reclamation, it is regarded as an invasive pest species by every country in which it grows, especially America and Australasia where, being highly inflammable, it presents a serious fire hazard. Once gorse becomes established, it is extremely difficult to eradicate; the plant has a lifespan of forty years and seeds can still germinate after lying dormant in the ground for thirty. Cutting encourages regrowth from the stump and burning simply creates aggressive seed dispersal as heat opens the pods. Glyphosphate herbicides will knock it back, but at the expense of killing off more sensitive plants in the area.

Even in its unmanaged state, gorse has considerable benefits in creating valuable habitat for a whole range of biodiversity. The extended flowering season attracts a mass of insect life, particularly rare butterflies and moths. The Sinnington Hunt manages a stand of gorse at Hutton Common which is one of the few places where the pearl bordered fritillary butterfly and the common dog violet on which the caterpillar feeds can be found. The dense thorny cover provides nestling shelter to a number of little birds; stonechats, linnets, wrens, rare Dartford warblers and yellowhammers, known in Wales as 'melyn yr eithin' – the yellow bird of the gorse. Apart from anything else though, a stand of gorse is always a good place for finding a fox.

Bats

My workshop, or glory hole, is in what was the feed room for the carthorses that were stabled in this part of the old farm steading. The gable end of the building faces south and in the loft above, is a pathetic little garret reached by a ladder, where a succession of single farm workers once lived. In 1923, one of them painted his name on the door and, more in hope than expectation, the word 'Welcome'. Behind the plaster and lath walls, with their shreds of flower-patterned wallpaper, a colony of tiny pipistrelle bats – the smallest in Britain – have established a nursery colony.

From late August, until the drop in temperature forces them into hibernation towards the end of October, male pipistrelles, gorged on the insects they have eaten through the summer, parade outside their breeding sites. This might be a crack in a wall, a hole in a tree or the gap under a roof tile, which the male vigorously defends. A love song, inaudible to human ears and wafts of pheromones attract a succession of females which are taken into the den for mating. Female bats have developed a system of delayed fertilisation, retaining the males' sperm separate from their eggs which enables them to avoid ovulating, if they are in poor condition when they emerge from hibernation in the spring. Bats hibernate from October until March or whenever the temperature drops below and rises above 9 degrees Celsius. In May or June, pregnant bats create maternity roosts in traditional sites, whilst males and barren females roost elsewhere. Gestation is around five to seven weeks and controlled by the weather. Warmth is essential to young bat pups and gestation can be arrested if there is a prolonged period of cold and wet.

Bats roost hanging upside down by their toes, but to give birth, the female hangs by her thumbs. She draws her legs up to create a pouch of her tail membrane into which the bald, blind pup is born in June or July. The tiny infant with its disproportionately large head and feet crawls up the fur on the dam's chest, latching on to her breast with its teeth and for the first week, or until the pup becomes too heavy, the dam continues to hunt with her infant clinging to a nipple. A bat's hooked teeth are thought to be specifically designed to provide the young with a secure attachment whilst the mother is in flight. Once the pups weight affects her flying ability it is detached, placed in a crèche of other infants and fed every couple of hours, until it is old enough to fly at three weeks. In about six to eight weeks, they are fully weaned and able to forage for themselves. This is the period I love. A stream of adult and juvenile bats, dropping like leaves out of a crack in the wall below the garret roof, just as the light fades and performing their spectacular aerial ballet, swooping and diving around the square formed by the farm buildings.

Our part of the Borders has a high rainfall and the boggy, misty countryside with its big areas of woodland, is ideally suited to the night-active insect life on which bats feed; midges, mosquitoes, lacewings, moths and beetles. We are particularly well situated for a variety of bat species – the stand of old larches, ash trees and Scots pines on the knoll behind the farm buildings provide roosts for noctule bats, one of our largest with a wingspan of 400 mm, which skim like swallows above the treetops. As dusk turns to darkness, common long-eared bats, Natterer's and whiskered bats flicker through the nearby ruins of Bothwell's gaunt castle, whilst scores of Daubenton's bats circle among the alder trees to dive for midges above the surface of the Hermitage Water. Not everyone is as lucky. The sixteen species of bat found in Britain are among our most endangered animals. Some, like the Bechstein's, barbastelle, serotine and greater horseshoe are considered under threat and now only found in the south and west. The heavy, slow-moving mouse-eared bat is probably extinct and rare sightings in Kent and Sussex are almost certainly vagrants from France.

The bat population, which had been in gradual decline since the 1900s, accelerated during the 1960s and 1970s, as intensive farming destroyed much of the habitat that provided their food and shelter. Thousands of hectares of insect-rich permanent pasture were ploughed out and wetlands drained. Pesticides further reduced the available food source and created health problems through bats eating contaminated insects. Hundreds of miles of hedgerows were grubbed out under field expansion schemes and acres of old deciduous woodland cut down, to be replaced by close planted conifers. Many bats relocated to old traditional farm buildings when they were replaced by modern units, only to be dislodged as these became popular for conversion. Increasingly, bats sought new roosts in the roofs of houses and there are an estimated 50,000 homes today providing accommodation to a colony of bats.

All bats became protected under the Wildlife and Countryside Act of 1981 and it is an offence to kill, injure or disturb them. At about this time, county bat conservation organisations, known as Bat Groups, started to be formed to increase knowledge, understanding and acceptance of these unique creatures. In 1992, the Bat Conservation Trust was created to act as an umbrella organisation to the growing number of county Bat Groups.

There is now a network of nearly 100 Bat Groups throughout Britain, with some 1,500 bat workers who provide rescue centres for injured bats. They survey roosts, monitor hibernation sites and advise householders, builders, farmers and foresters on how to preserve bat habitats. One of their most important roles is raising public awareness by organising monthly events and bat walks to visit known roosts from

May to September. Since bat activity takes place in rapidly fading light, the principal tool on these occasions is the bat detector. A bat detector is to the chiroptologist — as we bat aficionados are known — what binoculars are to the birdwatcher and is the ideal present for the person who has everything.

Bats are able to fly and locate prey by echolocation. Sounds emitted through the mouth or nose at incredible speed — up to thirty a second — are reflected back from objects, enabling bats to build up sound pictures in their brain of their immediate surroundings and to hunt, even in complete darkness. Bats burn an enormous amount of energy flying and their consumption of insects is prodigious — a tiny pipistrelle will eat 3,000 midges a night. Bat detectors convert these ultrasonic calls into sounds audible to humans, allowing us to plot movements and identify species through a fascinating series of clicks, ticks and smacks, which rise to a crescendo as they home in for the kill. Added to these, are extraordinary little bird like chirrups and squeaks detected during the breeding season.

Since 1997, the last weekend in August is European Bat Weekend; a celebration of bats and bat conservation, arranged by the Bat Conservation Trust. A whole range of walks, talks and public events are organised across Britain and throughout Europe, as many countries on the Continent who, like the UK, are part of the conservation group, Eurobats, set up light traps to draw in insects and bat detectors to record their resident bats as they swoop in to feed. A wonderful opportunity for anyone interested to join in the fun and learn something about the world's only true flying mammal.

Hound Trailing

A sport that, although synonymous to Cumbria, also takes place in the Western Borders of Scotland, Southern Ireland and, to a lesser extent, North Yorkshire, is hound trailing. In simplistic terms, hound trailing is a speed and endurance test during which specially bred hounds follow an aniseed trail of around 10 miles over ground varying from the steepest rock-strewn fell to boggy mosses, in every sort of weather from sleet to blistering heat. Hound trailing originated among the walkers of trencher-fed hounds from local fell packs, looking for a bit of out-of-season fun. At the end of the hunting season, the hounds from the fell hound packs would be dispersed amongst the farming community who would look after a hound or maybe a couple of hounds until the beginning of the next season.

Quite when the idea started of laying a trail and betting on the outcome of two or more hounds racing round it is anyone's guess. John Coughlan, who reported hound trailing events for many years in the *Whitehaven News*, mentions in his fascinating book on trailing that an entry in a Wensleydale parish record, dated 1750, refers to a villager being paid to lay a trail using a dead cat, but I suspect the sport had been on the go long before then. Its popularity spread, particularly among the traditionally hard-betting West Cumberland mining communities during the nineteenth century, and hounds began to be selected for speed and stamina purely for racing. Soon trailing was a regular feature of life in north-west England and the Borders north of Carlisle from early spring until late autumn.

As the sport became more competitive, some outcrossing to pointer, harrier and greyhound – even collie – was tried, in order to reduce weight and improve speed. A modern trail hound stands between 50 and 65 cm at the shoulder and weighs 13 to 25 kg, depending on whether they are a dog or a bitch. They are recognised as a true breed with the vital characteristic for any trail hound – drive, dash, nose and stamina – deriving from their fell hound ancestry. To preserve these qualities, an outcross to a foxhound dog is still allowed under Hound Trailing Association rules. By 1900 there were a number of hound trailing associations and no clear rules. To survive, the sport obviously needed regulating and in 1905 the situation came to a head when a hound that had been clipped down to the skin to reduce temperature was entered for a trail for the first time. The extra advantage in having been clipped allowed the hound to win its race, causing a riot among competitors and spectators, as opposing factions argued over whether this was cheating.

Robert Jefferson from Whitehaven, a man of considerable stature and local reputation, is credited with forming the Cumberland Hound Trailing Association in 1906, which continues as the governing body for the seven area committees in Cumbria. The Borders, Southern Ireland and North Yorkshire have their own governing bodies, but they all race under the same rules at international meetings. The HTA, with around 1,000 members, hold meetings every day of the week, the Border, with a membership of 150, has five or so a week and in North Yorkshire, one. There are different grades of trail at each meeting: 'senior', for hounds over two years old; 'puppy', for those over one year but under two; 'maiden', for hounds that have never won a championship trail and 'restricted', for hounds that have not won more than three trails in the current and preceding season. There are also 'veteran' trails for hounds over six. Entry fees are tiny – £1 per hound – and the prize money is correspondingly small, but all associations donate their proceeds after expenses to local charities, and with the economy of scale this soon mounts up.

The HTA have various main events through the season; the May Day Borrowdale trails, the Bitch and Dog Produce trails in June, the Festival of Hound Trailing in July, the August Premier and October HTA trails. All four associations are a feature of agricultural, game and county shows in their areas as well as, for the HTA, the famous Grasmere and Ambleside Sports and for the Borders, the Langholm Common Riding. The four associations meet annually to hold an international trail and all information is available through the internet or the *Whitehaven News*, West Cumberland's famous weekly newspaper, which devotes an entire page to the winners of the previous week's Hound Trailing Association meetings, hound form and the forthcoming fixture lists.

Nearly all hounds are owner trained and a close affinity between hound and trainer is vital. Although hound trailing is essentially a betting sport, with bookies present at every meeting, competitors give the impression of running hounds for the sheer pleasure of watching what is virtually the family pet racing. This is reciprocated by the hounds and it is quite obvious at any meeting that the fundamental reason they race with such uninhibited enthusiasm is to please their owner. Most people probably have three hounds at different ages in training and whilst exercise is crucial, infinite care goes into their diets, with many people possessing jealously guarded recipes passed down from generation to generation. A hound might be trailing three times a week, so getting the balance right between carbohydrates for stamina, protein for an energy boost and red or white meat to control the hound's temperature, depending on hot or cold weather, is a highly skilled art.

The trails are laid by two extremely fit trailers who clamber out to the halfway point of the designated route, carrying the trail cloths and a quantity of the scent mixture in which they are soaked – a blend of paraffin and aniseed made up by an officially appointed chemist. The trailers then walk away from each other, dragging the soaked rags, one going towards the start and the other towards the finish, so in effect the trail is more or less circular, with the start and finish always near the actual trail field where everybody parks their vehicles. There are scenes of high excitement when the trailers come tottering in and entries for first race, baying with excitement and straining at their leashes, are taken to the start line. Coats are stripped off to reveal incredibly lean foxhound types, most of whom have been trace-clipped down to the skin, all except the last 10 cm of tail, which is left to protect the tip. An official then dabs each hound on the shoulder with a coloured marker to thwart any attempt at race fixing by introducing a fresh hound halfway round – a dodge not unheard of in the bad old days. Handlers remove collars and grip a fistful of loose skin at the scruff; the start judge drops his white flag and the hounds are slipped, moving off in a pack which quickly changes shape as lead hounds pull ahead.

A senior trail must last a minimum of 25 minutes, or it is deemed that hounds have cut the trail, to a maximum of 45; puppy trails are roughly half the distance and half the time. The pack is often quickly lost from sight as it disappears over the skyline and spectators eagerly scan the horizon through binoculars as they wait for the hounds to come back into view. There is a buzz of excitement as the leaders are identified and a flurry of activity round the bookies' stalls as punters adjust bets, which can be taken throughout the trail. A distant line of trail hounds moving inexorably towards you down a distant fell side is an incredibly thrilling sight, particularly when you've got a fiver on the favourite, but it takes an aficionado to know one hound from another while they are still a long way off. With minutes to go, a video camera is set up to record the winner and owners carrying plastic or metal buckets containing titbits move down to the finish line. When hounds become clearly visible, a terrific cacophony of noise starts as owners call hounds in, shouting their name, blowing whistles and waving or banging feed buckets. Hounds are caught as they cross the finish line – an easy enough task as their heads are straight into the buckets – rugged up and checked for injury. Winners receive their 'ticket' and organisers immediately prepare for the next race.

Most of the hound trailers whom I know have inherited their passion for the sport, and it is this that gives hound trailing meetings such a delightful 'family outing' atmosphere. Membership spans the ages and there are any number of extremely healthy-looking septuagenarians – which is not surprising, considering a hound in training needs exercising about 5 miles a day – who have been trailing since they were at school. Their children have followed them into the sport and their grandchildren can be seen happily kicking a football round the car-park field whilst the family pet is racing.

Hound trailing quickly gets under the skin; the scenery, whether Solway coast, Dumfriesshire hill or Cumbrian fell, is always stunning, and as John Jacklin of the North Yorkshire HTA observed, 'like hunting, it takes you to places you wouldn't otherwise have gone to'. Even as a spectator, you become entranced after only a couple of meetings and I could listen to the craic of the older members all night long. Stories of laying trails with the effluent of road kills rotted down in an old milk churn, before aniseed came in during the fifties, or of young enthusiasts skiving off school to walk hounds miles to trail meetings back in the thirties. Tales of the hound nobbling and skulduggery that used to go on, or finding out where a trail was going to run and sneaking a hound round it before the meeting so that it was familiarised with the route. Getting a mate to lie out on the fell and signal where a particular hound was at a certain point in the race to swindle the bookies, or the enterprising chap who soaked the tip of a hound's tail in the urine of a bitch on heat to stop others passing it in a dog-hound-only race. Old countrymen reminiscing is part of our vanishing countryside, and for that alone a trail-hound meeting is worth a visit.

Ravens

The other morning, I was standing on the lawn idly watching four buzzards spiralling lazily in a thermal above the forbidding ruins of 'the guardhouse of the bloodiest valley in Britain', the Hermitage Castle. It was very peaceful; the buzzards were mewing contentedly, Tug the terrier dozed in a patch of sunshine and one of my bantam hens and her late brood of chicks fossicked for seeds in the long grass down by the stream. The tranquillity was short lived. A sudden sense of danger sent the hen into a fever of protective anxiety, hustling the chicks under a rhododendron bush and almost simultaneously, I heard the unmistakable sound of a raven, a cross between the satisfied grunt of a feeding pig and a leopard's sneering cough. With some difficulty I managed to locate a solitary bird soaring several hundred feet above the buzzards, revelling in his command of the skies by periodically performing a series of tumbling dives.

Ravens are possibly my favourite bird and their derisive double croak is a sound that never ceases to thrill me. Much of this has to do with their prominence as mocking harbingers of ill omen in mythology and early cultures of the northern hemisphere, which gave such enormous scope to childhood fantasies. The legend of their continuity at the Tower of London preserving the realm and the story of Churchill ordering young ravens from Wales, when the resident birds dropped to one during the Blitz, was heady stuff for any post-war schoolboy, but there was so much more. As a raven I could sail with the Norsemen in their war galleys to sack the East coast of Yorkshire or discover Greenland; skim across the centuries observing an endless succession of battles, or simply perch on the White Tower to watch the headsman at work.

By indulging my childhood fantasies, I learnt that this powerful bird, insouciantly sharing the same air space as golden eagles, were the largest and most intelligent of the crow family, which can be trained to hawk and readily learns to talk. They had an incredibly acute sense of smell, scenting sickness in animals and waiting for them to die. Eating virtually anything including hedgehogs they were voracious and determined hunters, known to go to ground in rabbit burrows after young or those of puffin, after eggs and fledglings. In hard winters they will dig through snow to reach buried sheep. In Greenland the flesh was once considered a delicacy and the islanders used the skins as under garments. Their wing feathers made the finest writing quills and were used for tuning harpsichords. In Bhutan, ravens are the national bird, representing one of that country's principal guardian deities. More than anything else, my affection for ravens stems from their presence in remote, wild places in circumstances that hold particular memories for me. The spectacular aerobatics of two ravens that entertained me as a child, when my parents fished on the Spey. The raven who congratulated me as I gralloched my first stag and the call of a raven that fills the silence when hounds check on a Cumbrian Fell.

Ravens once bred throughout the whole of Britain, with flocks of them feeding on filth in the streets, middens and around slaughterhouses in every town and village. Their role as scavengers was recognised as so important that they became protected during the Middle Ages, a status which lasted until sanitation improved and their food source became increasingly scarce. The urban population diminished, with the last pair of nesting ravens in London finally vacating Hyde Park in 1826. Over the centuries, they were gradually driven to the west and north as their habitat in our natural forests were destroyed and sheep grazings came under the plough, removing one of their carrion sources. By the end of the nineteenth century, they had disappeared from the Home Counties and virtually the whole middle and eastern side of Britain. Ravens briefly began to spread back during the war years and I suspect that massive conifer planting in the 1950s, myxomatosis and agricultural reclamations had as much to do with their retreat to the mountains, moorland and coastal cliffs of the west, as their persecution as a pest species. In 1970, the UK breeding population was estimated at 5,000 pairs, mainly in Wales and the western Highlands.

Since ravens became protected under the 1981 Wildlife and Countryside Act, the population has more than trebled and they are moving inexorably eastward into their former habitat. In the last few years breeding pairs have become established across the Midlands and south into the Home Counties. Malcolm Mercer, the National Gamekeepers Organisation chairman for Bedfordshire, Buckinghamshire and Hertfordshire tells me they are now at Woburn, Whipsnade and the RSPB headquarters at Sandy. Except for a pair that bred during the war, they have returned to Sussex for the first time since 1895. Nor are they confining themselves to the countryside. 'The devil in black feathers' is back nesting on Chester Cathedral after an absence of several centuries and a pair have taken up residence on a lighting gantry in the Liverpool docks. According to a recent article in the *Highbury and Islington Express*, a raven has been seen flying above Tufnell Park and Primrose Hill.

Like my bird, soaring above the Hermitage Castle, these sightings are an exciting novelty but it is a very different story for those who live near the epicentres of population expansion. Hen ravens lay between four and six eggs in February or early March. These hatch after three weeks and the young are flying six weeks later. They do not mate before their third season and until they pair up, young birds congregate in rookeries in ever-increasing numbers. It is these juvenile birds hunting in packs, which have recently created an ecological nightmare for every moorland gamekeeper I have spoken to. On grouse moors in west Aberdeenshire, sixty or seventy juveniles at a time are regularly seen during the nesting season, systematically quartering the ground on foot and eating the eggs and fledglings from every nest they can find. 'It is heartbreaking,' one of the keepers told me. 'No ground nester is safe and although we spend an enormous amount of time trying to drive them off, on a big moor they just move round you, or onto your neighbour's.' It is a similar story in Peebleshire, Northumberland, Yorkshire, Lancashire, the Peak District and the Lammermuirs, where there were none when I was farming in that part of the world a few years ago.

Sheep farmers on the Denbigh Moors in North Wales not far from three rookeries, one of which, Newborough Warren on Anglesey is reputedly the largest in Europe, regularly see 250 or 300 juvenile ravens towering in an updraft. 'They've devastated the bird life up here,' one of them told me, 'there used to be flocks of every sort of nesting wader in the spring – dunlin, golden plover, curlews, lapwings to say nothing of the grouse. The once familiar birdsong has all gone.' I asked him what it was like farming under these conditions. 'I wouldn't contemplate lambing outside nowadays; it would be murder.'

There has been a massive increase in all raptor species in the last twenty years and I was curious to know how ravens co-existed with them: 'They must watch the peregrines,' he told me. 'As soon as a peregrine kills, a raven materialises out of nowhere and drives it off. It's the same with golden eagles.' I remember being told that even eagles fear ravens and that Highland shepherds during the last century welcomed a nesting pair because they kept eagles away. Stories of ravens harassing various raptor species were repeated up and down the country, with one of the most telling coming from a county bird recorder for Greater Manchester. The BBC had asked him to monitor a peregrine's nest in a quarry near Bolton some years ago, to enable them to film the eggs at the point of hatch for the series *Britain Goes Wild* presented by the comedian, Bill Oddie. The recorder had just alerted the film crew that the peregrine chicks were hatching and Oddie was on his way, when a raven appeared and ate them.

It seems as if a faintly ludicrous balance of nature situation is developing, with an expanding raptor population predating on small bird species and ravens predating on raptors. I can't imagine what the various conservation bodies will make of this new turn of events, but I bet ravens have the last laugh.

Reading the Weather

Generations of countrymen used the saints' days of prediction and the old proverbs as a calendar around which to build the fabric of their lives, relying upon nature to provide them with immediate, localised weather forecasts on a day-to-day basis. When they left their homes in the early morning, they would instinctively look upwards and 'read the sky' in exactly the same way as their Neolithic ancestor would have done. Weather indicators, such as the colour of the sky, cloud formation, wind direction, scent, trees, plants, behaviour of animals, birds and insects, are as consistently trustworthy today as they have been for countless centuries. This information is depended upon by farmers, fishermen, gamekeepers, stalkers, hunt servants – in fact anyone whose lives are spent close to nature – but it is, of course, available to everyone. All they have to do is be observant and read the signs.

The direction in which the wind is blowing is one of the first signs a countryman looks for in the morning; he may deduce this from the movement of clouds, by pulling up a little grass and throwing it up into the air, or from the way the treetops are bending. An indication of the importance to rural people of wind direction is reflected in the sheer number of weathervanes on old farm buildings, often in the shape of a running fox, and weathercocks on church spires. Weathervanes – the word comes from the Saxon *fane,* meaning flag – have been in use since antiquity. Perhaps the earliest recorded is the one that adorned the Tower of Winds in Athens, built in the first century BC. As Christianity spread across Europe, weathervanes began to be erected on the top of church spires for the benefit of the community. Cockerels became the universal motif after a ninth-century papal edict commanded that every church in Christendom must be adorned with a cockerel, to remind

Christians of their duty of worship and not to betray Christ as St Peter had done. A reference to Christ's prediction at The Last Supper: 'I tell thee, Peter, the cock shall not crow this day, before that thou shall thrice deny that thou knowest me.' Although not originally intended as weathervanes, cockerels soon replaced any that already existed.

An indication of how important the knowledge of wind direction was in the Middle Ages, is reflected in one of the images in the eleventh-century Bayeux Tapestry, which depicts an artisan erecting a weathercock on the spire of Westminster Abbey. The first poem about wind to be printed in Britain was written by the farmer-poet, Thomas Tusser: 'Description of the Properties of Winds at All Times of the Year', published by Richard Tottel in 1557:

> *North winds send hail, south winds bring rain*
> *East winds bewail, west winds blow amain;*
> *Northeast is too cold, southeast not too warm,*
> *Northwest is too bold, southwest doth no harm.*
>
> *The north is a noyer to grass of all suites,*
> *The east is a destroyer to herd and all fruits;*
> *The south, with his showers, refresheth the corn,*
> *The west, to all flowers may not be forborne.*
>
> *The west, as a father, all goodness doth bring,*
> *The east, a forebearer no manner of thing;*
> *The south, as unkind, draweth sickness too near;*
> *The north, as a friend, maketh all again clear.*
>
> *With temperate wind, we be blessed of God;*
> *With tempest, we find, we are beat with His rod;*
> *All power, we know, to remain in His hand,*
> *However wind blow, by sea or by land.*

> *Though winds do rage, as winds were wood,*
> *And cause spring tide to raise great flood,*
> *And lofty ships leave anchor in mud,*
> *Bereaving many of life and of blood,*
> *Yet true it is as cow chews the cud,*
> *And trees, at spring to yield forth bud,*
> *Except wind stands, as never it stood,*
> *It is an ill wind turns none to good.*

The vagaries of wind and its consequences are just as significant today as they were to Tusser in sixteenth-century Tudor England. As a general rule, the prevailing wind in this country is from the west, and westerly winds usually predict temperate weather. This fisherman's proverb covers all the permutations of wind direction and their predictions:

> *When the wind is blowing in the north,*
> *No fisherman should set forth,*
> *When the wind is blowing in the east,*
> *'Tis not fit for man nor beast,*
> *When the wind is blowing in the south*
> *It brings the food over the fish's mouth,*
> *When the wind is blowing in the west,*
> *That is when the fishing's best!*

Northerly and easterly winds are associated with low atmospheric pressure and signify bad weather. Wind circulates clockwise round a high-pressure cell and anticlockwise round a low one. If the wind is out of the east, it indicates that a 'high' front from the west has just moved on or is passing to the north. Low pressure follows high with the same monotony as night following day, since highs and lows invariably alternate in progression. With the approach of a low weather front, the east wind picks up and becomes gusty. Known as the 'lazy wind', because it blows straight through you, easterlies are unpleasantly hot, dry and dusty in summer, and bitterly cold in winter. The north winds, which follow a low weather front, are cold, blustery and, from a sailor's point of view, always produce heavy, possibly dangerous seas. Westerly winds generally suggest milder weather, while a southerly in summer often brings humidity and rain: 'A wind from the south, has rain in her mouth'. This would explain the allusion to fish feeding in the proverb above, as insects would fly closer to the surface of the water in warm, damp weather.

> *No weather is ill, if the wind be still.*

Still, cloudless summer days are a sign of a dominant high-pressure cell; they can also signify the approach of a thunderstorm – the calm before the storm – as high pressure in the west sucks up surface wind before it can arrive locally. This type of storm, with its army of great towering clouds, threatening thunder, lightning and heavy rain, can easily be seen banking up to the west.

> *When the wind backs and the weather glass falls,*
> *Prepare yourself for gales and squalls.*

A backing wind is one that starts in the west and changes direction anticlockwise to the east or southeast, heralding the approach of a low-pressure cell and lousy weather.

Strewing Herbs

Wild flowers reach their peak in the heavy, airless days of July, when hedgerows, riverbanks and meadows are a riot of colour. Pink dog roses, yellow honeysuckle and bramble flowers twine through hedges, competing with greater bindweed and old man's beard. Verges are cloaked with magenta willow herb, lilac teasels and feathery-leafed yarrow, blue meadow cranesbill, yellow rattle, purple butterwort, pink bistort, foxgloves, daisy-like mayweed and red poppies. Old pastures burst with colour: fluffy yellow lady's bedstraw, orange bird's-foot trefoil, ox-eye daisies, purple betony and greater burnet, pink betony and knapweed, blue devil's-bit scabious, lilac spotted and southern marsh orchids, white mouse ear, white and pink clover, and the bumblebee's favourite, meadow vetchling. Water forget-me-not, water mint, gypsywort, brooklime and yellow-flowered flag irises thrive on the edges of streams and rivers. But the flower that is it at its best in July is meadowsweet.

Filipendula ulmaria grows in profusion in damp meadows, on boggy streambanks, pond and ditch sides throughout Britain from June to September, but the exquisite, heavy sweet scent from the foaming clusters of tightly packed, creamy-white flowers is particularly noticeable now. Both the flowers and the large, pinnate leaves are fragrant and these were once harvested as a strewing herb. Our ancestors were aware that ill health was associated with the 'noxious vapours' from the middens and open sewers that were part of life until the late nineteenth century. They believed that diseases carried by foul air could be avoided by superimposing a pleasant, and therefore healthy, smell over a revolting one – hence the reason plague doctors wore a beaked mask, that dread symbol of pestilence and death, which they filled with a potpourri of dried flowers, medicinal herbs and spices.

Well into the eighteenth century – for even longer in churches and poor households – the universal floor covering was rushes, which cushioned the feet on stone, wood or packed earth and absorbed spilt food and the rain, mud or snow brought into houses. Supplying the larger towns was a considerable industry, with rush cutters harvesting common rush, bulrush and the pleasant-smelling sweet flag from neighbouring marshes and delivering them by barge. Once laid, the rushes were sprinkled with 'strewing herbs', containing properties that were aromatic, disinfectant or vermin repellent. Among them were meadowsweet, a favourite of Queen Elizabeth I, for its fragrance; wormwood, catnip, pennyroyal, rue and fleabane as insect and rodent repellents; sage, thyme, rosemary, fennel, basil, burnet, mint, sweet gale, feverfew and lavender for their antibacterial and antiseptic properties. 'But those herbs which perfume the air most delightfully, not passed by all the rest, but being trodden upon and crushed, are three; that is, burnet, wild-thyme and water mints. Therefore, you are to set whole alleys of them to have the pleasure when you walk or tread', Francis Bacon, *Essays, Civil and Moral* (1601). Lavender was still being strewn on the floors of hospitals during the reign of Queen Victoria and was once extensively cultivated, with plantations in Kent, Suffolk, Surrey and London – hence Lavender Hill. The essential oils were released as the herbs were walked on and the grander houses employed someone whose role was to ensure the correct mixture of herbs and fresh flowers was strewn to keep the household sweet-smelling, healthy and free of insects and vermin.

When King Charles II returned from exile in 1660, he found the stench permeating St James's Palace from the Thames – which was little more than an open sewer – so disgusting, he created the office of King's Herb Woman or Royal Herb Strewer. Bridget Rumney was the first, paid the considerable sum of £12 annually as 'Garnisher and Trimmer of the Chapel, Presence and Privy Lodgings', plus an allowance for 2 yards of scarlet cloth for her livery. When the king married Catherine of Braganza in 1662, another £12 was paid for performing the same duty in the queen's apartments. The importance attached to the office is reflected in the role of the King's Herb Woman, dressed in scarlet livery, leading the coronation processions for a succession of monarchs over the following 150 years. The last full-time herb woman was Mary Rayner, who served the royal household from 1798 until 1836. Rayner, however, was considered too unprepossessing to lead the procession at the coronation extravaganza of George IV in 1823 and was replaced for the day by an imposing fifty-year-old matron Anne Fellowes. Subsequent monarchs dispensed with the ceremonial duties of the Royal Herb Strewer, although the office is still extant and is claimed today by Jessica Fellowes, a niece of the actor Julian.

Revered by the Druids as a sacred plant, meadowsweet once had a wide range of uses, including flavouring mead and sweetening sour wine. Decoctions made from the leaves were said to relieve diarrhoea, gout, rheumatism, tumours, blockages in the bladder and headaches, while a salve made from the flowers steeped in goose fat was put on burns. In 1838, salicylic acid, synthesised in the 1890s to make aspirin, was isolated from the plant. The word 'aspirin' is derived from 'spirin', based on meadowsweet's scientific name *Spiraea*. Nowadays, this wonderful plant seems woefully underutilised. The flowers can easily be used to make bath oils, cordials, wine or champagne, ice cream, jams, jellies, cakes or puddings. Meadowsweet fritters are exquisite and the addition of flowers to a rice pudding ensures that, for once, it really is ambrosial. As an enhancer, meadowsweet has an infinity of possibilities – I use it to flavour snuff.

Ancient Trees

Uniquely to Britain, we have amongst our fragments of remaining ancient woodlands, in churchyards, parks of great houses, on village greens and parish boundaries, a number of trees of immense antiquity, more than the whole of Europe put together. Since 2004, The Woodland Trust, Tree Register of the British Isles and the Ancient Tree Forum, have been compiling a register of heritage trees. Known as the Ancient Tree Hunt, volunteers are encouraged to find and map all ancient trees across Britain so that a comprehensive living database is created, enabling these remarkable specimens to be protected and preserved.

Yew trees of great age have survived in churchyards across the country, simply because they were there, in some cases for centuries, before the church itself was built. Yew trees have all the components that made them particularly attractive to the Druids; they are evergreen and produce red berries in winter, giving a splash of colour in a stark winter landscape where most other plant life was dead or dying, symbolising the enigmatic power of nature. Being poisonous and immensely long lived, they represented both death and immortality, with the Druids believing a grove of yews trees to be immensely powerful and using them as sites of pagan worship. That yews became synonymous with churchyards was due to the instructions given by Pope Gregory in AD 597 to the Benedictine monk Augustine, as he departed from Rome on his mission to convert the heathen Britons to Christianity. Gregory insisted that Augustine should not destroy places of pagan worship, but consecrate the sites to Christian use.

Virtually every churchyard has a yew tree and some are spectacular; the magnificent yew in St Bartholomew's Church, Much Marcle, Herefordshire, is 1,500 years old and has a girth of 10 metres. The 1,000-year-old yew which towers over the medieval church of St Peter's in Tandridge, Surrey, and the even older one in the churchyard of St Mary and St Peter, Wilmington, Sussex. The three great yews that dominate the little church of St Cuthbert's in Beltingham, Northumberland, are thought to be 2,000 years old, as is London's oldest tree, the yew in St Andrew's churchyard, Totteridge. The Ankerwycke Yew, close to the ruins of St Mary's Priory at Runnymede was already 1,200 years old when King John sat beneath its branches to sign Magna Carta and 1,500 years old when Henry VIII wooed Anne Boleyn at the same spot. For many years, the enormous yew tree which grows in the graveyard of the parish church at Fortingall in Glen Lyon, Perthshire, was believed to have been planted 3,000 years ago and the oldest living thing in Europe. This venerable tree was knocked off its perch when tree-dating experts calculated the yew in St Cynog's churchyard at Defynnog, a tiny village near Sennybridge, Powys, was at least 3,000 years old when Christ was born.

There are many more wonderful historic specimen trees, including the great limes of Holker Hall in Cumbria; the Maryculter sweet chestnut on the banks of the River Dee in Aberdeen, near the ruins of the thirteenth-century St Mary's Chapel, built by the Knights Templar and the great plane tree which towers over the grounds of the twelfth-century Mottisfont Abbey, near Romsey in Hampshire. The splendidly gnarled and ancient wild cherry tree in the grounds of Studley Royal in North Yorkshire; the Belvoir Oak, Ireland's oldest tree, in Belvoir Forest Park, Belfast, or the superb cedar of Lebanon at Addington Palace, Croydon, one of the Great Trees of London.

Broadleaved trees of immense age, particularly pedunculate and sessile oaks, have survived due to the archaic custom of pollarding and wood pasture, which was still practised in some parts of the country up until the eighteenth century. Pollarding evolved from coppicing – harvesting the new growth from the stump of a felled tree – as animals became domesticated. Instead of felling the tree close to the ground, they were cut at a height of between 2.5 and 3 metres which protected the regrowth from grazing animals in what became known as wood pastures. The reason we have such a remarkable heritage of historic trees, is because pollarding tends to make them live longer by maintaining their growth in a partially juvenile state, when it was most resistant to disease and by reducing the weight and windage of the top. Pollarding was particularly prevalent on common land where commoners had the right of 'estovers' and could harvest the leafy foliage, known as wood hay, as fodder for animals or for fuel, fencing and building material building. The Domesday Book records that nearly all woodland in Britain which was not wildwood or coppice was some type of wood-pasture.

During the Middle Ages, pollarding was a feature of Royal Forests, chases and deer parks of the nobility, clergy and major landowners. Many of these contained trees which were already venerable in the seventeenth and eighteenth centuries when park designers such as Kent, Bridgeman, Capability Brown and Repton absorbed them into the landscape, to give an air of antiquity to the new parks they designed to complement great houses. Windsor Great Park has many gnarled, bulbous and stumpy pollarded beech and oaks of great antiquity, with their tangled 'stags head' tops, including the Conqueror Oak, reputedly planted by Duke William himself. The park at Blenheim Palace has the largest collection of historic pollarded oaks in Europe and Scotland has the ancient sessile oaks in Cadzow Park near Glasgow, once home to a herd of rare White Park cattle.

There is the 750-year-old Royal Oak of Richmond Park; the Old Man of Calke Park in Derbyshire is 250 years older and the one at Bowthorpe Park Farm, in Lincolnshire, with a girth of 12 metres, is older still. Among the oaks in Sherwood Forest is the Major Oak, with a girth of 18 metres and a hollow interior large enough to hold fifteen men or thirty small children, famous for the Victorian legend that Robin Hood used it as a hideout. Savernake Forest in Wilshire has a wonderful concentration of veteran beeches and oaks, including the 1,200-year-old Big Belly Oak, whilst Epping, Hainault and Hatfield forests, all that remain of the Royal Forest of Essex are internationally famous for their historic pollarded beeches and hornbeams.

These are just a fraction of our priceless legacy of historic trees, and it never ceases to amaze me that we have so many ancient broadleaves due to the discovery of pollarding by Neolithic pastoralists at least 2,000 years before the birth of Christ.

Kit

In the corner of my hall is an old, elephant-foot stick stand overflowing with walking sticks. There are some with attachments for swatting thistles or digging out docks; ram's-horn leg or neck crooks and half a dozen home-made hazel-shanked thumbsticks; a silver-topped Malacca sword stick with a Toledo blade; a spiked Austrian bergstock and an African knobkerrie bound with copper wire; a bamboo cane with a spirit level and boxwood measuring rod inside for measuring the height of a horse, made by Brigg; a selection with animal-head handles in bone, ebony or ivory – a terrier, greyhound, fox hound, horse – and one made of ivy with a snake entwined along its length.

Each has a special place in my heart, but my favourite is an old, battered wading pole, with a hand piece made from the forked tines of a stag antler, a lead-weighted antler stub at the bottom and a leather strap with a spring clip, attached to a link whipped onto the hazel shank. All I know of the stick's provenance, is that it was made for my grandfather sometime before the Second World War, by a ghillie on the Delfur beat of the Lower Spey called Colin and it is a credit to the stickmaker's art that it has survived intact for so long.

A purist might say this stick lacks the aesthetic beauty and natural elegance of a true hazel thumbstick and they would be right; the shank is too thick, the weighted stub on the bottom makes it too heavy, the hand piece is too chunky and away from the river, the leather strap with its spring clip and link looks completely out of place. However, apart from the purpose for which it was originally made, this ugly duckling of the stick world has been a true and faithful companion. It has freed me from the viscous ooze of countless tidal mudflats on 'fowling expeditions; hauled me up Cumbrian Fells when I have been walking-out; has been strong enough to drag a beast off the hill and then, there is the detachable leather strap. This was made of ¼-inch bridle leather and measures 3½ feet long with a loop to go over the shoulder. Obviously, its original function was to stop the stick being carried away whilst casting in fast-flowing deep water, but it is astonishing how useful the strap is on terra firma when one needs two hands and there is nowhere convenient nearby to prop the stick against – using a pair of binoculars, for example, or lighting one's pipe.

Practicalities aside, the reason the stick is my favourite bit of kit, is because it is above all, comfortable to use and that is essential in a stick. The hand piece must have come from an old beast who had been doing himself well, hence it is proud and well sprung, with a hollow below the V of the tines into which the pad at the base of one's thumb fits snugly. Although heavy, the weighted stub makes the stick perfectly balanced and aids locomotion, but more importantly, it is exactly the right height. The role of a thumbstick is to propel one over rough ground and to take the weight off the body when standing for any length of time. Thumbstick aficionados fall into two distinct groups: leaners and loungers. Leaners can be seen propped upright with arms folded, leaning insouciantly on the V of their stick; loungers place it in their oxter and lean. I fall into the latter category.

Exotica

The seventeenth, eighteenth and nineteenth centuries were the era of great landscape gardeners and as Britain became a colonial power, exotic plants from all over the world were introduced to Britain. John Tradescant the Elder and his son were both gardeners to King Charles I and introduced the horse chestnut, scarlet runner beans, larch trees, apricots, Virginia creepers, yucca plants, tulip trees, pitcher plants, bald cypress trees, magnolias, phlox and asters. In the early part of the eighteenth century, geniuses such as John Claudius Loudon, William Kent, Stephen Switzer, Charles Bridgeman and Henry Wise created magnificent gardens and stunning landscaped parkland at Windsor Castle and Kensington Palace, St James's Park and Hyde Park, Chelsea Hospital, Longleat, Chatsworth, Castle Howard, Blenheim Palace, Chiswick House, Cliveden, Rousham and Stowe, to name only a few. From 1719 at Rousham, in Oxfordshire, for example, Charles Bridgeman and William Kent created a vast Augustan landscape in a curve of the River Cherwell to recall the glories and atmosphere of ancient Rome. Paths wound through woods and little groves, where water from the Cherwell was diverted to create small rills leading to larger ponds and formal pools. Classical statuary of Roman gods and mythological creatures were cunningly positioned to catch the eye as paths led from cascades to water gardens and on to the next temple or arcade, each set in its own valley or glade, creating a string of picturesque tableaux.

It was whilst Bridgeman and Kent were transforming the 162-hectare baroque park at Stowe, in Buckinghamshire, filled with avenues, lakes, temples and pavilions designed by the architects James Gibbs and Giacomo Leoni, that Lancelot 'Capability' Brown was taken on as a pupil. Brown was arguably the most prolific and famous British landscape designer of his time, creating around 170 parks surrounding some of the finest country houses, most of which have endured to this day and are open to the public. His style focused on perfecting nature in huge landscape parks carved out of the adjacent countryside. Formal gardens were replaced by great vistas of smooth undulating grass running straight to the mansion house; serpentine lakes formed by invisibly damming small rivers and clumps, belts, groves or scattering of trees judiciously positioned to accentuate a curvature of the ground or highlight the skyline.

Humphrey Repton was the last of that generation, designing parkland and gardens for nearly fifty stately homes, most notably at Stoneleigh Abbey, Blaise Castle, Welbeck Abbey, and Woburn Abbey, Russell Square and Endsleigh, for the Duke of Bedford. Repton specialised in creating picturesque landscapes; at Endsleigh, the Duke of Bedford's fishing lodge on the Tamar in Devon, which I remember well from the days in the 1960s when my grandparents took the fishing, Repton created a fantasy world of many secret gardens. The house, a magnificent *cottage órne,* was built to designs drawn up by Sir Francis Wyatt between 1811 and 1814 on a bluff overlooking the Tamar Valley across to the thickly wooded Cornish bank. Repton improved on the breathtaking natural beauty of the position by creating rose walks and terraces that lead to summer houses and grottoes, hidden dells or crags with viewing seats. Acres of lawns tumble down to the river, past lily ponds, cascades, a Gothic garden and fernery, a hollow filled with giant gunnera, a miniature ice house, an octagonal dairy, a shell grotto, and a Holy well. Behind the house is a stunning arboretum of exotic specimen trees, chosen to create a wonderful combination of colours; Himalayan birches, Japanese cedars, weeping beeches, Persian ironwoods, tiger-tail spruces and Douglas fir.

David Douglas, after whom the tree was dedicated, was one of a number of intrepid eighteenth- and nineteenth-century plant hunters who risked their lives in the search of exotic plants for botanical gardens or wealthy patrons. Douglas undertook a plant-hunting expedition in the Pacific Northwest in 1824 that ranks among the great botanical explorations of a heroic generation. On his return in 1827, he brought back 240 new plant specimens, including Sitka spruce, sugar pine, Western white pine, ponderosa pine, lodgepole pine, Monterey pine, grand fir, noble fir and several other conifers that were to transform the British landscape and timber industry, as well as numerous garden shrubs and herbs such as the flowering currant salal, lupin, penstemon and California poppy. In 1834, whilst searching for plants on Mauna Kea in Hawaii, he toppled into a deep pit dug by the natives to trap feral cattle and, whilst waiting to be rescued, was crushed to death when a wild bull fell in on top of him.

Douglas was among an extraordinary group of intrepid men who became known as plant hunters, who risked their lives in the hazardous occupation of searching for rare plants, to satisfy the growing passion for exotica in Britain. One of the earliest was William Kerr, a Scottish plant hunter, who was sent to the Far East by Sir Joseph Banks and spent ten years from 1804 exploring China, Java and the Philippines, sending back 240 new species, including the rose *Rosa banksiae*, or Lady Banks' Rose, named after his patron. In 1812, Banks sent him to Ceylon (now Sri Lanka), where he died of opium addiction in 1814 'unable to prosecute his work in consequence of some evil habits he had contracted, as unfortunate as they were new to him'.

In 1840, James Veitch, the entrepreneur seedsman catering to the demand for exotic plants, sent William Lobb to South America with instructions to send back anything interesting he could find and in particular, seeds of the fabled *Araucaria araucana,* or Chilean pines, later to be known as monkey puzzle trees. *Araucaria araucana,* had first been brought to England in 1795 by Archibald Menzies, the naval surgeon and botanist, who had grown a handful of seedlings from pips on the voyage back from Chile and Veitch believed these would be hot sellers.

Two years later, racked with fever from plant hunting in the Amazon basin, William Lobb struggled through snowdrifts to reach a stand of Chilean pines growing at 2,000 metres up on exposed ridges of volcanic peaks in the southern Andes. Too weak to climb the trees, Lobb managed to collect over 3,000 seeds by shooting down cones with his rifle. The monkey puzzle seeds arrived safely in England, whereas many of the other plants that Lobb collected during four years of exploration in the steaming jungles of the Amazon or freezing foothills of the Andes, rotted in the corner of a warehouse in Guayaquil, overlooked by the shipping agent who had promised to see them safely on a boat to England.

Undeterred, Lobb returned to the interior between 1845 and 1848, successfully sending home a variety of hardy and semi-hardy trees and shrubs, among which were the Patagonian cypress 'Prince Albert's yew', Chilean totara and Antarctic beech. After a flying visit back to Britain to confer with James Veitch, who was making a fortune growing seedlings from Lobb's discoveries, he was off again in 1849. This time to the St Lucia Mountains of southern California, sending back seeds of the St Lucia fir, the California buckeye, the bishop pine, gray pine, coulter pine, knobcone pine, the Californian yew, a miniature horse chestnut and many shrubs and flowering plants, most quite new to

British gardens. Moving north, he collected seeds from the Western white pine and the California redwood, the Colorado white fir, the red fir, the Western red cedar and a gigantic tree which was simply called the monarch of the Californian forest, until renamed Wellingtonia, after the Duke of Wellington.

By now Veitch's business had grown to the extent that he was able to open grand premises in London's Kings Road and, despite Lobb showing obvious signs of ill health in 1854 – he had contracted pretty well every tropical fever known to medical science during the previous fifteen years and was beginning to exhibit signs of syphilitic monomania – Veitch nevertheless packed him off again to southern California. As the illness progressed his behaviour became increasingly erratic and the few new discoveries he was able to make, such as a previously unknown variety of white fir and the rare Torrey pine, were sent to Veitch's rival, Low's Nursery in Clapham. Veitch was apoplectic and remonstrated furiously, to which Lobb responded by sending any interesting live plants and herbarium specimens he found direct to Sir William Hooker, the director of the Royal Botanic Gardens at Kew. By 1860 all contact with Lobb was lost – Veitch liked to think he had absconded to the California gold mines, when in fact, he had become so ill that he was reduced to vagrancy, dying in 1864, forgotten and alone in the paupers' ward of St Mary's Hospital in San Francisco.

There were many others including John Fraser, William Lobb's brother, Thomas, who also collected for Veitch's Nursery and, until he lost a leg in the Philippines, specialised in rare orchids from the Far East. Joseph Hooker, who introduced twenty-eight species of rhododendrons from an expedition to the Himalayas in 1850, fuelling the Victorian craze for these shrubs. John Gould Veitch, one of the first plant collectors to be allowed into Japan, brought back Japanese maples, the sawara cypress and other conifers, as well as palm trees from the Philippines, before dying of tuberculosis aged thirty-one. Charles Maries, who spent four years in China and Japan between 1876 and 1880, collecting hundreds of rare plants amongst which were Japanese oaks and conifers, rhododendrons, Chinese and fire dance witch hazels, dwarf bamboo plants and golden larch. Maries subsequently became superintendent of the gardens of the Maharajah of Durbhungah, where he laid out the very extensive grounds which surrounded the palaces, before spending the remainder of his life as Superintendent of the palace gardens and parks belonging to the Maharajah Scindia of Gwalior.

The last of the great nineteenth-century pioneer plant hunters, whose bravery gave us some of the most remarkable trees and the loveliest plants ever grown, was Ernest 'Chinese' Wilson. In twenty years, from his first expedition in 1899, Wilson introduced a range of over 2,000 rare Asian species at a time when it was generally believed that every worthwhile plant in the Far East had already been discovered. His first expedition was to the isolated valleys in the Hupeh Province of China, where he spent two years collecting some 906 different species, including paperbark maples, kiwi fruit trees, Julian's barberry, various genii of clematis, the rare dove tree, dwarf holly, primrose jasmine and candelabra primulas. In 1903, Wilson returned to China and discovered the priceless, exquisitely scented regal lily growing beside the Min River in a remote valley of Western Szechuan Province, introducing it successively to Britain. Wilson also collected plants for the famous Arnold Arboretum of Harvard University in Boston and was twice sent back to the Min River by Charles Sargent, the director, who desperately wanted regal lily bulbs for the arboretum. On the first expedition in 1908, the ones Wilson found all rotted en route to Boston and on the second, in 1910, he and his porters were caught in an avalanche of boulders as they toiled along the Min Valley. In true nineteenth-century explorer style, Wilson was being carried in a sedan chair at the time and was unable to extricate himself before it was hit by a large rock. His leg was badly crushed and, after setting the broken bones with the tripod of his camera, he was strapped to a litter and lugged back to civilisation in excruciating agony on a forced march lasting three days. Thereafter, he walked with a pronounced limp, but it was bulbs from this expedition that successfully introduced the regal lily into cultivation in the United States.

Terriers

At the Game Fair in 2018, I was wandering down Gunmakers' Row when I found myself on a collision course with a stout lady cradling a small, bright-eyed, smooth-haired, black and tan dog, with a long tapering muzzle and pricked lugs, not unlike a miniature Dobermann pinscher. In fact, as the good lady was quick to tell me, it was an English Toy terrier, known in America as the Toy Manchester, since it is regarded as a variety of the same breed as the standard size. Manchesters evolved in the early Industrial Revolution when a whippet, or 'snap dog', was crossed with an old English black and tan terrier to create a dog that excelled both at ratting and coursing rabbits. Her cherished companion would be a direct descendant of a dog called Tiny the Wonder, who in 1848 and 1849, at the height of rat-baiting popularity, held the record for killing 200 rats in under an hour. Tiny weighed only 5½ pounds and can be seen in action in the painting *Rat-Catching at the Blue Anchor Tavern, Bunhill Row, Finsbury*, which hangs in the Museum of London. I felt this was probably not the moment to tell the owner she was cuddling a rat killer of such illustrious lineage, but it did make me wonder how many people have any idea what all the different breeds of terrier were actually bred for.

Until the late 1800s, terriers were loosely classified as four basic types: long-legged, wire-coated and probably black and tan dogs of the Welsh hill country and parts of Ireland; white dogs in the south and west country; rough-coated dogs north of Edinburgh known collectively as Scotch terriers; whilst Northumberland, Cumberland and the Scottish Borders had predominantly red or brown flat-coated terriers. There were, of course, an infinity of variations among these types to conform with personal preference or their intended use: badger, otter, or fox. Each district would have a terrain and quarry species which governed the type of terrier bred there – short legs and hard to ground might suit in one area, where a longer-legged type would be more appropriate in another. Geographical isolation led to a preponderance of line breeding and the only opportunity for an outcross was meeting someone with a likely looking terrier at the little local markets, or if a seasonal labourer or passing gypsy happened to have one.

Increased mobility during the Industrial Revolution, when roads improved and railways opened the country up, led to agricultural shows and great improvements in all livestock, as the best in one area were selected to improve those in another. Dog shows became phenomenally popular after the first, purely for setters and pointers, was organised by the gunmaker, W. R. Pape and held at the annual Newcastle Cattle Sale in 1859, with J. H. Walsh, editor of *The Field,* as one of the judges. Mixed breed class shows followed at Birmingham and London in 1860 and 1861; the Kennel Club was formed in 1873 and by the time Charles Cruft held his 'Great Terrier Show' at the Royal Agricultural Hall, Islington, in 1886, with 57 classes and 600 entries, many of the old, amorphous regional working terrier types were becoming established as distinct breeds. Among them were the following.

Skye & Cairn

Until the nineteenth century, when selective breeding for type and colour became fashionable, the highly regarded Scotch terriers were described as being of two sorts: one, a very strong type, with a long coat, short legs and elongated back, bred principally for bolting otter and badger, which became known as Skye terriers. The other sort were, 'rough coated and beautifully formed, having a shortened body and more sprightly appearance'. Among these is the Cairn, known as the short-haired Skye terrier until the breed was renamed in 1909, bred for bolting foxes from among the cairns where Highland foxes tend to hole up and for general vermin control, which included everything from a rat to a wildcat. They were quite bold enough to take on an otter; in the 1890s, Captain Macdonald of Waternish in north-west Skye, kept a pack of forty Cairns which he regularly used for bolting otters from cairns formed by fallen rocks along the seashore below the cliffs at Waternish Point.

White West Highland

Light-coloured or off-white Scotch terriers had been around for centuries before King James I sent six from Argyllshire to France, as a gift for Henry III. Several of Edwin Landseer's paintings of the early nineteenth century portray light-coloured Scotch terriers, especially *Dignity and Impudence* depicting the head of a pure white terrier sharing a kennel with a bloodhound, which perfectly captures the pricked lugs and keen, alert expression of these little dogs. A close cousin of the Cairn and sharing all their pluck and intelligence, we have Colonel Edward Malcolm, 16th Laird of Poltalloch, to thank for standardising the breed. Legend has it that Malcolm was so horrified at accidently shooting one of his brown terriers, which he mistook for a hare at his estate near Lochgilphead, that he determined never to make the same mistake again by breeding pure white ones.

Scottish

Originally known as the Aberdeenshire terrier and bred for bolting fox, badger and otter, among other Highlands vermin, these immensely hardy, determined terriers were more like a Skye terrier, with their powerful jaw, low-slung bodies and short, sturdy legs, than their close relatives the Cairns or White West Highland. Captain Gordon Murray was responsible for standardising the breed in 1879 and Scotties were to become immensely popular after James Buchanan, the whisky magnate, marketed a brand of whisky with a label depicting a White West Highland and a Scottie. Famous owners of Scotties include Queen Victoria, Presidents Roosevelt, Eisenhower and Bush, Jacqueline Kennedy Onassis, Rudyard Kipling, G. K. Chesterton and Hitler's girlfriend, Eva Braun.

Dandie Dinmonts

With their low-set, long, flexible body, shaggy coat and stumpy legs, Dandie Dinmonts are one of the historic terrier breeds of the Scottish Borders. Bred by the sporting farmers, shepherds and gypsies for bolting badger and otter, they were said to fear nothing with a hairy skin. Now a very rare breed, the Dandie Dinmont of today is believed to descend from dogs owned by a gypsy known as 'Piper Allan' in the latter decades of the eighteenth century whose terriers were renowned across Northumberland. Lord Ravensworth offered him 50 guineas for one after Allan cleared otters from the lake at Elsingham Park, and the Earl of Northumberland is said to have offered him a rent-free farm for another, both of which he refused. The breed is named after the character in Sir Walter Scott's novel *Guy Mannering* (1815), based on the sporting farmer, James Davison of Hyndlee in Roxburghshire, who had several terriers of Allan's breeding. Descriptions of them in Scott's novel led to their immediate popularity, with the Dandie Dinmont Terrier Club being formed in 1875 and the breed standard created by William Wardlaw Reid.

Border Terriers

Border terriers are probably the most popular pet among the terrier breeds, with their broad skull, short muzzle, slightly protuberant eyes and flat, wiry coat. A close relative to the Dandie Dinmont, they were referred to as the Coquetdale or Redesdale terrier from the area of the Cheviot Hills where they were most commonly found. Bred for bolting foxes they were quite courageous enough to take on badger and otter, but canny with it and knew when to keep out of trouble. From the late 1800s they became known as Border terriers from their long association with the hunt of the same name. The Robson and Dodd families, who had owned and hunted the hounds for several generations, were famous for breeding hardy, working terriers, long enough in the leg to follow hounds. Borders were recognised by the Kennel Club in 1920 and are one of the few breeds to have remained true to type.

Bedlingtons

These curly coated, leggy terriers with their high, rounded skulls and long, tapering muzzles, share a common ancestry with both the Border and Dandie Dinmont, with a whippet cross somewhere in their genealogy. They were the archetypal general-purpose vermin and rabbiting dog, much favoured by the 'Border muggers', as the gypsies based at Yetholm were known, as well as Border farmers. James Davison of Hyndlee was among those who kept a pack of terriers made up of what were to become known as Dandie Dinmonts, for otter and badger; Border terriers for fox; and Bedlingtons, that had the reputation for fighting anything their own weight to the death and for running down whatever bolted. Originally known as Rothbury terriers, Bedlingtons became very popular for racing and rabbiting with the miners of the Northumbrian coalfields in the early nineteenth century, hence the name.

Sealyhams

These were bred from 1850 until his death in 1891 by Captain John Tucker-Edwardes of Sealyham House in Pembrokeshire, principally for bolting otter and badger. Tucker-Edwardes was dissatisfied with his existing 'mongrel' terriers and set about breeding his ideal: a small, strong, brave, active terrier, with short legs, a harsh white weatherproof coat and a strong jaw. He kept no records of the mix, but it is believed he used corgi for length of back, White West Highland from his friend Colonel Malcolm for colour and the now extinct Cheshire terrier, a small type of bull terrier, for breadth of jaw and courage. Tucker-Edwardes' terriers became well known and highly regarded for their pluck and intelligence; they were officially recorded by the Kennel Club as Sealyhams in 1911. By 1935, the show dogs were described in *Hutchinson's Popular & Illustrated Dog Encyclopaedia* as: 'much too large and much too clumsy for the work they were originally bred for, and they would not have the ghost of a chance of getting even their heads into an otter's holt'. Mercifully, enough of the working gene survived for Harry Parsons to form the Working Sealyham Terrier Club in 2008. Harry has worked tirelessly to maintain this endangered breed and regular hunts rats with a pack of seventeen or so.

Norwich

These immensely sporting little dogs and their close cousins, the drop-lugged Norfolk terrier, are the smallest working terrier breed and what they lack in size they make up by being extremely active, vocal, self-opinioned and bobbery. Reputably bred from local East Anglian 'mongrel' terriers and small, red, Irish terriers belonging to itinerant seasonal labourers, with possibly Cairn blood, they are great ratters and were often kept for controlling vermin in barns and stables. Known at one time as Cantab terriers, from their popularity with the university students, or Trumpington terriers, after the livery stables in Trumpington Street, where James Barrons the ostler, bred a particularly sporting type. Both prick- and drop-lugged Norwich terriers were accepted by the Kennel Club in 1932, and in 1964 the drop-lugged type were reclassified as Norfolk terriers.

Patterdales

A particularly hardy type of terrier known collectively as fell terriers, had been bred in Cumbria for centuries to run with the fell foot packs of the Lake District, where foxes have always been the scourge of the Lakeland sheep farmers. A fell terrier needed to be long enough in the leg to run all day with hounds, narrow in the chest to follow a fox when he went to ground in the rocky borrans, and tough enough to ensure that if he didn't bolt, he wasn't left alive. In 1873, the Matterdale and Patterdale hunts amalgamated to form the Ullswater and, in 1879, Joe Bowman became huntsman, a position he held pretty well continually for the next forty-five years. Bowman bred old-fashioned working Border terriers, but used an outcross to a local fell terrier now known as the Lakeland, to breed a terrier type he named the Lakeland Patterdale, after the valley he was born in. Patterdales became famous after Brian Nuttall began breeding them in the 1960s from Bowman's line, which had been bred true to type by Cyril Breay and Frank Buck before the Second World War.

Airedales

The largest of the terrier breeds, Airedales, with their tight, weatherproof coats, were bred by crossing the old black and tan terrier of the north of England with an otterhound, to create a general-purpose dog that, although too big to go to ground, would be needle sharp and determined at anything above. Once known as the Bingley terrier, this cross proved to be incredibly clever and adaptable, with a good nose for following scent, capable of being broken to a gun and able to retrieve, both from land and water. Their unique characteristics were quickly recognised by Lieutenant Colonel E. H. Richardson, a specialist in dog handling and training, who provided Airedales for the Russian Army in 1904 to act as message carriers and to locate the wounded during the Russo–Japanese war. From 1908 he supplied Airedales to the North Eastern Railway police for patrolling dock areas. Airedales were extensively used during the First World War by the military and Red Cross, and the many stories of their bravery and stoicism under fire brought them to public attention. They became very popular in the 1920s and 1930s but are now rather rare.

Jack Russells

None of the short-bodied, stumpy legged, little dogs, known as Jack Russells, full of mischief and *joie de vivre*, would have been recognised by the famous Devonshire sporting parson as having any resemblance to the dogs he bred. Russell described his foundation bitch, Trump, as 'white with a patch of dark tan over each eye and ear while a similar dot, not larger than a penny piece, marks the root of the tail. The coat, which is thick, close and a trifle wiry, is well-calculated to protect the body from wet and cold but has no affinity with the long rough jacket of the Scotch Terrier. The legs are straight as arrows, the feet perfect, the loins and whole frame are indicative of hardiness and endurance, while the size and height of the animal may be compared to that of a full-grown vixen fox.'

Russell wanted terriers which could run with hounds, were bold enough to bolt a fox or lie with it whilst it was dug out. Essentially, he wanted terriers that were sensible enough not to get hurt and he prided himself on being able to say that none of his ever had. The Parson terrier, finally recognised by the Kennel Club in 1990, is probably nearer to the original, or the Fox terrier – Russell was a founding member of the Kennel Club and helped write the breed standard for the Fox terrier. He never showed his own breed and on the day of his funeral in 1883, all Russell's breeding notes and records were last seen blowing about the yard of his house at Swimbridge. It is highly unlikely any terriers today can claim to be a descendant of Russell's, but there any number in hunt kennels of whom Russell would have said, 'Now they are the right sort.'

Autumn

Hanging Game

For many years, BASC has held an annual sporting auction to raise funds to support its Young Shots Training Days. Among the many and varied sporting opportunities on offer, there has been, for the last fifteen years, the chance to bag, 'Three days guided wildfowling with Sir Johnny Scott on the National Nature Reserve at Lindisfarne for two or three guns, accommodation included.' Lindisfarne is one of the most beautiful places in Britain and the great tidal dish created by Holy Island's elongated shape and the curved elbow of the Northumbrian coastline provides a protected habitat for the largest congregation of waders in Europe. The trip, organised by BASC Northern Region and the Newcastle Wildfowlers Association, is intended to provide the winning bidders with the ultimate wildfowling experience and the opportunity of a crack at the 14,000 or so wigeon that migrate to Lindisfarne during September from the Baltic and around 8,000 Icelandic pink-footed geese which begin arriving at about the same time on their staging post to East Anglia.

Last year's trip took place during the last week in October and although the dawn flight was a blank, as skein after skein of pinks lifting off their shore roosts with a thrilling clamour, swung away well out of range, the evening flight produced a bag of eleven wigeon. That night seven of us sat down to dinner in the dog-friendly Lindisfarne Inn: me, Kenn Ball and Paul Scott of the Newcastle Wildfowlers Association, who were our wildfowling guides, Colin Teago, the BASC Wildfowling Warden for the National Nature Reserve, and the auction winners; Giles Cadman, Malcolm Newbert and Robert Fathers. All were experienced game shots, loved eating it and in the case of Giles, an enthusiastic game cook. During the course of the evening, the conversation touched on the length of time any of those present hung game to allow muscles to relax and flavour to develop, with the general consensus of opinion, regardless of species, that no longer than two or three days at about 4 degrees Celsius was necessary.

The Countryside Alliance Game to Eat Campaign suggests five to seven days, depending on the weather and, of course, personal preference, although the shorter two-to-three-day view is supported by Annette Woolcock, the development manager of BASC's Taste of Game initiative, as does Lee Maycock, the chairman of the Craft Guild of Chefs, and William Alldis, The Shotgun Chef. William has his own shoot and restaurant, The Cart Shed, on the family farm near Epping, an outside catering business and can provide shooting lodge cooks trained to handle game properly. All agree that game hung for only a few days provide an infinity of culinary options, which would not be appropriate to anything hung for longer.

Fifty years ago, it would have been unthinkable to eat game other than well-hung. Denys Watkins-Pitchford, the great naturalist, sportsman and author, better known by his pen name 'BB', recommended four days for duck and eleven for geese – even an old gander, but that was really quite a revolutionary short hanging period compared to most popular opinion at the time. In the 1968 edition of Florence White's *Good Things in England*, for example, she recommends Major Hugh B. C. Pollard's *The Sportsman's Cookery Book*, as the best all round cookery book for game. The major, a splendid Edwardian character, author, firearms expert and one-time member of MI6, who listed his hobbies in *Who's Who*, as hunting and shooting, recommended hanging pheasant for three weeks or 'until their tail feathers give', whilst partridge or grouse should be hung as long as possible. Annette Hope, in *A Caledonian Feast* (1989), mentions a butcher of her acquaintance in the 1960s, whose customer judged a grouse ready to eat when maggots could be heard moving inside it.

Eating game high had been in fashion for centuries – Colonel Peter Hawker, the father of wildfowling, whose prowess as a shot whilst serving with the 14th Light Dragoons during the Peninsular Wars, earned him the soubriquet of 'Wellington's Honorary Wildfowling Officer' by the members of the duke's staff, invented a sauce which can only have been intended to disguise the flavour of very high game. This consisted of port, lemon juice, lemon rind, chopped shallots, pounded mace, coarse red pepper, strong vinegar, tarragon, thyme, brandy, grated horseradish and a healthy dollop of mushroom catsup. Hawker's mushroom catsup or ketchup, bore no resemblance to anything on sale today and was more a loose paste made from salted mushrooms boiled with mace, pepper and vinegar.

Professor John Wilson, writing under the pseudonym 'Christopher North', was a principal contributor to *Blackwood's Magazine* from 1820 to 1835, and recommended a game sauce made with port wine, salt, caster sugar, lemon juice, mushroom catsup, coarse cayenne pepper – the quantity to be doubled if the game was particularly high – and a generous helping of Harvey's Sauce. Harvey's Sauce, which became a household name and was originally the invention of the mother of Captain Charles Combers, was simply known in the family as Rotten Fish Sauce and made from fermented anchovies, vinegar, garlic, Indian soy sauce, cayenne pepper, mushroom ketchup, with an added dash of cochineal to colour it red and make it look less unattractive. Combers was a Meltonian, known for his breathtaking performance across country as the 'Flying Cucumber', in the days when Hugo Meynell was Master of the Quorn (1753–1800). Combers, who never travelled without a bottle of his mother's sauce, was in the habit of stopping for the night on his way to Leicestershire at the George Inn at Bedford, run by a man called Harvey, who had previously been chef to the Duke of Bolton. Harvey was so impressed by Combers' sauce that he acquired the recipe, refined it, made it commercially and went on to make a fortune from it.

Obviously, there were endless occasions when game sent by cart, carrier and even rail, once the railway system pushed through to Inverness in 1850, would be humming by the time it reached its destination and the happy recipients would take the view that game was meant to be eaten high. And yet, there were and always had been, plenty of opportunities to eat game fresh – birds shot or trapped in the coverts of big estates, long before the days of driven shooting. Duck decoys were common on inland waterways, to say nothing of the waterfowl shot by the market gunners to supply local coastal towns. Hawker alone, once he was invalided out of the army after the Battle of Talavera in 1809, spent virtually every waking moment in the pursuit of edible bird life, all of which could have been eaten fresh, or at least, not in a state of near decomposition. Even allowing for the jaded palates of the Georgians and their heavy consumption of sweet, fortified wines – port, Madeira, malmsey, marsala, the dark sweet sherry, Jerez Dulce and the cherry liqueur maraschino (the Prince Regent once sent a naval vessel from Malta to Zadar on the Dalmation coast to collect 100 cases), there seems little reason to eat all game high, drenched in tongue-shrivelling sauces.

Nor did the Victorians and Edwardians change their attitude to hanging game and eating it high; the main culinary difference was in replacing the strong, spicy, pepper sauces of the previous generation with fruit. Oranges, plums, prunes, raisins, apples, redcurrant jelly, cherries, quinces, bananas – even in the 1920s, Boodle's was still serving grouse and partridge stuffed with mashed bananas – and in Scotland, juniper berries. A popular sauce for game north of the Border was made with shallots, garlic, sugar, thyme, claret and crushed juniper berries, simmered in game stock, whilst the berries were often used as stuffing and in game pies. Cumberland sauce, both hot and cold, actually an adaption of a recipe of Hannah Glasse's dating back to 1747, was very popular and made with port, the juice and zests of a lemon and an orange, dry mustard, redcurrant jelly and red pepper, to which Auguste Escoffier, the great French chef, added ginger. One of the best and simplest was invented by Elmé Francatelli, the first celebrity chef, who worked at various times for Queen Victoria, the Prince of Wales at Marlborough House, Crockfords and the Reform Club. His sauce was made from two tablespoons of port added to half a pound of redcurrant jelly, a bruised stick of cinnamon and the thinly paired rind of a lemon.

It will have been common knowledge from the dawn of time that the muscles of an animal stressed at the point of slaughter, particularly a bird driven to rapid flight, will be tense and therefore, require a period of hanging to allow the meat to relax, tenderise and for flavour to develop. However, it does seem extraordinary that it has taken so long for people to realise that game need not be eaten ripe and that there appear to be no records, at least that I can find, of anyone bucking the trend. Victorian game larders were specifically designed to be cool, dry, places and once the game laws were relaxed in 1832 and game shooting rapidly increased with the developments of the breech loader and establishment of sporting estates, game became much more available. At one time, beams of game hung in Leadenhall Market and outside every provincial butcher's shop during the season, and one would have thought surely, an invalid or someone requiring a less rich diet, would have discovered that game did not need to be eaten in a state of virtual decay. Nevertheless, the belief persisted century after century, that the longer game was hung, the meat became more tender and the flavour improved. As we say in the Borders, it was a classic case of 'it's aye been', which no one, until recently, thought to change.

We have modern chefs and people such as Peter Barham, the author of *The Science of Cooking*, for our improved knowledge and thank heavens for them. Plucking game a few days old is immeasurably easier and a great deal more pleasant than anything that has been hung for longer and I remember the joy of shooting my first cock pheasant, almost as vividly as being given it to pluck three weeks later when it stank to high heaven and the skin tore with virtually every feather. Gutting it when the job was finally finished, is something I prefer to forget.

A Mast Year

Nature seems to be tugging the last of summer's goodness and energy out of the land in October, as she draws up her defences and prepares for the trials and tribulations of the coming winter. There is almost a sense of defiance resonating in the glorious colours of autumn; the gold, bronze, russet and ochre of leaves turning on our deciduous trees, and the yellows, purples, reds or scarlet of the last hedgerow berries. Although not as good as last year, nature has been kind to us again; we have made enough damson vodka and sloe gin to last the season, and the kitchen reeks of jam and hot vinegar, but then, 2013 was a mast year and an exceptional one at that.

The word 'mast' in botany, is derived from the early Germanic *maesten*, meaning to fatten or feed, and is defined by *Encyclopaedia Britannica*, as 'the nuts or fruits of trees and shrubs, such as beechnuts, acorns and berries, that accumulate on the forest floor, providing forage for game animals and swine.' A mast year is a natural phenomenon in which certain trees, such as ash, chestnut, English oak, field maple, beech, hawthorn, hazel, hornbeam and Scots pine, produce a glut of seeds compared to almost none in others; beech, for example, typically produce a mast year every five to ten years. One of the mysteries of the natural world is exactly why mast years occur and as Simon Toomer, National Specialist for Plant Conservation at the National Trust and previously director of the Forestry Commission's National Arboretum at Westonbirt observed: 'Part of the fascination of experiencing a mast year is that we don't completely understand the complex blend of factors that give rise to them and allow plants to coordinate the production of so much fruit and seed.' On one level, there is clearly a degree of reproductive protection at play, with individual species periodically producing a disproportionately large quantity which exceeds the needs of seed predators. Known as 'predator satiation', this ensures that enough seed survive hungry mouths to germinate and become seedlings the following spring. On another level, climate plays a large part and after a prolonged period of hard, wet weather, nature seems to step in to create a synchronised glut of mast, enabling desperate wildlife to replenish itself and build up energy levels for procreation, with enough seed left over for future stocks of trees and shrubs.

The autumn of 2013 was a classic example; it had started to rain the previous February and continued with barely a dry day right through the catastrophic spring and summer of 2012. The ground was sodden by May and compared with previous years, there was a visible decline in flowering plants, trees or shrubs and noticeably less insect activity to aid pollination. Major country shows such as Badminton, the Scottish Game Fair and the Great Yorkshire at Harrogate were cancelled one after the other right through the season, to say nothing of dozen upon dozen smaller shows and by the end of August, the summer was recorded as the dullest, coldest and wettest for a hundred years. In September, the Woodland Trust monitors reported the worst year for fruiting since they began keeping records in 2001, among them all deciduous trees, particularly beech and oak, hazel; brambles, rose hips, haws, rowans, sloes, crab apples, damsons and holly.

This dearth of autumn fruit and nuts posed a desperate situation for wildlife hoping to lay on fat or store food for the winter. In the wild, ponies, cattle, pigs and all UK deer eat acorns, conkers, beech mast and sweet chestnuts, which is one of the reasons these trees were always planted in deer parks. Red and grey squirrels, jays, wood pigeons, woodpeckers, dormice, wood mice, pheasants, chaffinches and nuthatches are among those who variously depend on acorns, hazelnuts and sweet chestnuts. Hedgehogs, dormice, foxes, badgers and swarms of insects all love blackberries, and hundreds of little songbirds rely on autumn hedgerow fruits for their survival, not least of which are the flocks of hungry redwings and waxwings driven south by freezing temperatures in the subarctic.

After a long, wet summer, it would not have been unreasonable to expect a dry frosty winter, but it was not to be. Month after month ground by with a wearying succession of bitter winds, rain, sleet or snow. In late February the glass dropped to minus 18 Celsius and on 23 March, a blizzard howling in from the Atlantic blanketed much of the country, burying sheep, tearing down power lines and trapping motorists across the north of England, including seventy-two who had to be rescued from vehicles buried under 5-metre drifts on the A595 near Millom in Cumbria. Was this the last vicious sting in the tail of a wretched winter, before the green shoots of spring? Unfortunately not; biting winds, rain and snow flurries were a feature of April, the month when one expects to hear birdsong and see regrowth, as high pressure blocked the progress of the jet stream which normally brings warmer weather. Our hill lambing at the end of the month was the worst I can remember; despite tons of extra feed, ewes were lambing and walking away from their lambs. There was none of the jubilant territorial song of the spring migrants who nest in the uplands, snipe, curlew or lapwings and the only birdlife seemed to be the ever-watchful, black, sinister ravens and carrion crows – what a feast they had.

By the second week of May there was no sign of improvement; snow still lay on the high ground, trees and shrubs were bare of buds, heather brown and lifeless, not even a hint of hedgerow blossom or wildflower growth and the few swallows that arrived soon left for want of insects to feed on. Suddenly, it was as if nature had decided enough was enough and almost overnight, everything changed. The temperature rose, deciduous trees and shrubs acquired a green halo of buds, wild flowers bloomed, insect life erupted, the swallows returned and the long-awaited cacophony of birdsong – the billing and cooing of fond pursuing – was glorious to hear. Conditions remained absolutely perfect for flowering, pollination and seeding throughout the rest of May, June and July with gradually increasing temperatures, plenty of sunshine, no extremes of wind and just enough rain to stop the ground drying out or inhibit insects from pollinating. Strangely enough, although it hardly seemed possible there could have been any benefits from the appalling weather of the previous twelve months, the incredibly heavy rainfall was a key element in the process of photosynthesis, which creates the energy needed for reproduction in the form of starch and sugar. Furthermore, the cold, late spring meant that many spring-flowering trees and shrubs, such as hawthorn, oak or rowan among many others, delayed flowering until the weather improved and were now fruiting more closely together as nature lifted the lid on pent-up growth, rather than their normal staggered fashion. Unless we had gales and heavy rain in August, all the indications pointed towards a mast year of epic proportions.

Nor did August disappoint; warm and humid, the little rain that fell only helped to fill fruit and by the end of the month, it was clear the predictions were right. In a few weeks, there was going to be a bumper crop of acorns, beech mast chestnuts and conkers. Hedgerows were already dripping with fruit and gleaming with the autumn colours of ripening blackberries, wild raspberries, rose hips, haws, elderberries, sloes, damsons and crab apples, whilst out on the moor, a branch on the rowan tree which grows by the ruins of an old shepherd's bothy, broke under the weight of the scarlet berries. Wildlife had a bonanza, with a constant stream of birds from dawn to dusk, from wood pigeons to little warblers such as blackcaps and whitethroats, gorging on nature's bounty. Clouds of bees, wasps, butterflies and hoverflies fed on the fallen fruit and in the evenings an incredible number of different moths, providing a feast for bats laying on fat before the temperature dropped and they retreated into hibernation. The glut of hazelnuts were ripe to pick a week earlier than their traditional date of 14 September, Holy Cross Day and the Verderers' Court of the New Forest announced that due to the enormous crop of acorns, beech mast and chestnuts, the date of the pannage season would be brought forward to 9 September and run until 15 December.

Pannage, or common of mast, is the historic right of those living on common land or in Royal Forests, 'commoners', to fatten domestic pigs for an agreed period in the autumn. Mast feeding pigs was once widespread across England and the mainstay of pork production, but is now virtually restricted to the New Forest National Park. The season usually starts in the third week of September and runs for a minimum period of sixty days, although the exact dates are decided by the Court of Verderers, whose role is to protect and administer the New Forest's unique agricultural commoning practices and the Forestry Commission, which manages the forest on behalf of the Crown.

Grazing pigs in the Forest is an important part of the ecology and their essential role is to reduce the risk in a glut year to the livestock of those with rights of common of pasture, the right to graze horned cattle, ponies and donkeys, from dying of liver and kidney damage caused by gorging on acorns. The acorn crop was so high in 2013 that by mid-November, 50 of the 5,000 or so ponies had died and 16 of the 3,000 cattle that roam the forest. From medieval times until the end of the nineteenth century as many as 6,000 pigs were turned out every autumn, but exercising the right of pannage has declined with the escalating prices of smallholdings and properties in the forest, and the influx of new owners with no desire to keep pigs. Sadly, nowadays the number of pigs on the forest during the pannage season, each with its identifying ear tag and nose rings to stop them rooting, fluctuates between 250 and around 500 in a mast year, with a handful of 'privileged' sows allowed out at other times, provided they are taken back to the commoner's holding at night. Such a shame, the flavour of a ham made from pannage pork is akin to the best Spanish *jamón ibérico de bellota*, but twice as rare.

Dryland Huskies

On the second weekend in October, the early morning tranquillity of the Ae Forest in Dumfriesshire was broken by the arrival of a fleet of long wheel-based Transit vans and the unusual sound of ecstatic vulpine howls. Since before light, around 200 Siberian huskies and their handlers had been preparing for the opening dryland sled dog rally of the season, organised by Ann Shaw, Vice Chairman of the Scottish Siberian Husky Club. At 8.25, a purpose-built, three-wheeled tricycle without a seat, weighing 12 kg and known as a monkey rig or suicide cart, was pushed to the start point. Six huskies wearing fitted body harnesses and bouncing on the end of leads were brought into position on either side of a 10-metre central gang line, attached to the rig's front pillar.

With the first musher (from the French, *marche*) poised behind the rig, huskies hooked to the gangline and held steady by assistants, the timekeeper, helpers and spectators start a five-second countdown. On the word 'go', the huskies fling their weight forward and the rig shoots off followed by the sprinting musher. Once the huskies have settled into their pace, he leaps onto footplates at the back of the rig and is off on a roller-coaster ride along twisting forest tracks at an average speed of 32 kph. Control is entirely dependent on the dog's intelligence and their response to a handful of simple commands: 'gee', for right; 'haw', for left; 'hike', for speed up and 'easy', for slow down.

As this was the first meeting and the glass unseasonably high – huskies cannot perform in temperatures over 12 degrees Celsius – the course was limited to 4.3 km. At other times, UK course lengths are roughly

1.6 km per dog in the team. At the Ae Forest, over eighty competitors from Scotland and as far south as the Midlands had the choice of classes ranging from six, four, three and two husky races. With one- and two-dog events for juniors aged from eight to eleven and twelve to fifteen. There were also single-dog scooter races; Canicross, where competitors are attached to their dog by a rope and run behind it, plus heats for a small number of Samoyed and Alaskan Malamute enthusiasts.

Sled dog racing is one of the fastest-growing winter sports with a fan base stretching from Alaska to New Zealand, regardless of whether arctic conditions normally associated with dog sledding exist. Over 30,000 participants compete annually in around 2,000 race meetings worldwide. Its origins and the gene pool from which all Siberian huskies in Britain stem, goes back to the Alaskan gold rush at the end of the nineteenth century. In 1898, gold was discovered at Anvil Creek on the remote and desolate Seward Peninsula of north-western Alaska, attracting a flood of prospectors from the Klondike gold fields. This turned to a stampede when gold dust was found among the sand on the frozen, storm-lashed beaches of Cape Nome. By the early part of the century, Nome became the largest habitation in Alaska as 20,000 miners, investors, gamblers and get-rich hopefuls – including the legendary Wyatt Earp – established a sprawling tent and shack city along the barren coastline.

With subarctic weather conditions for much of the year, every human commodity from mail to mining equipment had to be hauled in from the nearest port of Anchorage by hundreds dog sled teams. Large dogs of any breed were stolen from all over America to meet the demand and the big, heavy boned freighting huskies of the indigenous Mahlemut Eskimos fetched fantastic sums. With so many dogs in Nome and not a lot to do, it was only a matter of time before bored miners started betting on the speed of individual teams. These became weekly events, culminating in the annual 657-km All Alaska Sweepstake for a purse of $10,000.

In 1909, Walter Goosak, a Russian fur trader entered a team of small, alert-looking dogs from Siberia. At around 20 kg and 56 cm at the shoulder, these were half the size of the Malamute and mongrel sled dog competitors and were greeted with open derision. To everyone's astonishment, these 'Russian Rats' came third. Their performance was enough to convince the Hon. C. F. M. Ramsay, a younger son of the 13th Earl of Dalhousie, who was in Nome overseeing the family's mining investments, that here was a winning breed.

That summer, when the ice on the Bering Straits melted, he chartered a schooner and sailed up the Anadyr River to the trading settlement of Markovo in eastern Siberia. Over a period of weeks, he purchased sixty of the best specimens he could find from the nomadic Chukchi tribesmen. In 1910, he entered three teams of Siberian huskies in the Alaska Sweepstake, driving one himself. Between them they came first, second and fourth. The winning team, driven by 'Iron Man' Johnson, set a record of 74 hours 14.5 minutes which has never been beaten. Ramsay's Siberian huskies attained enormous admiration as racing dogs and fast freighters – a working husky will pull ten times its own body weight – and their progeny became national heroes in 1925, at the hands of the famous breeder and musher, Leonhard Seppala.

In January that year, Nome was stricken by a rapidly spreading epidemic of diphtheria. Isolated by snow and pack ice, the population risked annihilation unless vital anti-toxins could be brought in. Planes were ruled out because of the extreme cold and Nome's only hope lay with the huskies. Serum was sent from Anchorage to the nearest rail head at Nenana, 1,078 km from Nome. Twenty-two of the most-experienced mushers volunteered to face temperatures of minus 30 degrees and blizzard conditions to relay the precious medicine through. Leonhard Seppala and a team of twenty Siberian huskies travelled over 419 km in appalling weather to his place in the relay and a further 145 km carrying the serum. Following the 'Mercy Run', as it became known, Seppala and his Siberians toured the northern states of America, initiating a passion for huskies and recreational dog sled racing.

Wheeled husky racing started in Britain after a pair of Siberian puppies were acquired by Don and Liz Leich in 1969. As the dogs developed, their daughter Sally recognised that the only way they could be properly exercised, was in the manner for which they had been bred. Dog sledders in New England used wheeled rigs for summer training and Sally acquired a set of instructions and built one. The idea caught on amongst the handful of other husky owners and before long, they were meeting to stage little races. Popularity for the emergent sport had grown sufficiently by 1978, for the newly formed Siberian Husky Club of Great Britain to stage the first organised wheeled husky race meeting. There are now at least seven sled dog clubs or associations, with over 400 dog teams competing at more than sixty rallies, during a season running from October to the end of March. These cater for Kennel Club registered breeds – Siberian, Malamute, Samoyed, Eskimo and Greenland, plus open classes for Scandinavian hounds: a relatively new breed of pointer cross husky, which are gaining an international reputation.

Nearly all racing and training in Britain takes place on Forestry Commission land – twelve forests in Scotland and about thirty in England issue licences to clubs. The Ae Forest is the first of two dedicated training and racing sites in the country. Designed and created by Steve Lindsay, the project manager of Dog Sport Scotland and twice winner of the Dry Land Sled Dog World Championship, its aim is to utilise the sport to increase visitors to Dumfries and Galloway. It is an indication of the potential of this initiative, that among the sponsors are The Forestry Commission, Scottish Enterprise, Scottish Power, Barony College, local government agencies and Buccleuch estates, who host the UK Dog Sled Final at Drumlanrig Castle.

There is no shortage of help and encouragement for anyone interested in taking up the sport and each club runs beginners' classes. Ann Shaw recommends starting with a welfare dog – all associations operate a welfare scheme for homeless huskies – before acquiring a pup. Siberians are the most popular breed and considered the best natured. They are highly intelligent, mischievous, easy going, gregarious and very good with people of all ages. In their natural environment, they would have lived with the Chukchi, keeping the children warm at night.

However, huskies have been bred for a specific purpose and their overwhelming desire is to run. When not on leads, they need to be kept in a secure exercise enclosure, at least 1.8 metres high. A husky that gets loose will simply run until he is lost and the last sight a musher has if he falls off his rig, is the team disappearing into the middle distance. They retain the hunting instincts of their wolf ancestors and cannot be trusted with other animals. One of a musher's anxieties is ground game crossing a team's path. To suddenly find oneself hurtling into the undergrowth at 20 mph can be a memorably unpleasant experience.

Dog sledding easily becomes obsessive. 'We used to have a city-centre flat and an Audi TT,' Steve Holmes and Lisa Curtin told me. 'Ate in smart restaurants and went abroad twice a year. Then we fancied a dog to go with the lifestyle and bought a husky. Now, we live miles from anywhere. The Audi has been replaced by a Transit van and we have eleven huskies living in what was the garden. We train three times a week and compete virtually every weekend. Sometimes we wonder how it happened, but you can't beat it.'

Elder

Elder (*Sambucas negra*) is a common sight in hedgerows, woodland edges, waste ground and the sites of previous habitation. From late October, the emaciated branches of an elder tree are the epitome of winter's starkness: bare, gaunt, brittle and lifeless. In spring it comes alive with new growth and from June, when the plant is covered in profusions of creamy white, sweet-smelling blossom, it is more synonymous of the British summer than any other vegetation. 'Summer is not here until the elder flowers and it is gone when the berries ripen,' is an old country saying with a lot of truth to it.

The tree has an enormous number of medicinal, culinary and domestic uses as well as a wealth of folklore and mysticism attached to it. One of its names, Judas tree, refers to a belief that Judas hanged himself from the branches of an elder tree. Another myth was that the cross on which Christ was crucified had been made of elder wood and it was considered bad luck to use elder branches on a fire. Neither are very likely; elder is technically a bush, with fragile pithy branches which would never provide enough wood for a cross or be strong enough to bear the weight of a hanged man. The deeply entrenched belief that the elder had the power of protection against evil – in some parts of Britain a corpse would be buried with a piece of elder wood and the driver of a hearse carried a symbolic whip made of elder – predates Christianity and is descended from pagan plant worship. It was frequently planted near houses both to ward off witches and warlocks and to utilise the leaves, flowers and berries. Digging near elder trees is often rewarding as ash and household rubbish was spread on the ground to fertilise and nurture it. I have found all sorts of treasures near the one that grows by the farmhouse – bits of plate, old bottles, coins, earthenware pots and once, a child's tiny hobnailed boot with a turned-up toe.

In the spring the young leaves can be made into a purée, eaten as spinach or, with the addition of flower buds, as a delicious salad. But it is in June when the shrub is in blossom that it really comes into its own. There are probably more uses for elder blossom than any other species of flowering plant. A refreshing summer drink, reminiscent of vanilla soda can be made, simply by putting some flower heads in a jug and pouring boiling water over them. Add yeast and sugar to make a pleasantly alcoholic elderflower 'champagne'. Boil flower heads with water and sugar to make a cordial for bottling; capture the sweet scent in custard, compotes, jellies, sorbets and ice cream, pancakes and doughnuts; mix with honey, wine and a little garlic as a glaze for roast lamb. Elder flowers are even wonderful on their own, in a sandwich of thinly sliced brown bread. My absolute favourite pudding is elderflower fitters; whole flower heads dipped in light batter and deep fried. After a hot tiring day, elderflower heads are as good in a bath as meadowsweet.

At the end of August, the flower heads die back to be replaced in a few weeks by clusters of brilliantly versatile reddish-black berries. These can be used as a fruit in soufflés, tarts, pies, jellies and sorbets, cakes and muffins or as an exquisite sauce for game, particularly venison or duck. They can also be simmered with stock to make an incredibly rich soup, boiled with sugar to make jam or a syrupy cordial similar to crème de cassis, or pickled with vinegar to make a sweet and sour sauce for cold meat or roast lamb. We sometimes make spicy Pontack sauce, which was popular among the nineteenth-century hunting set, who based themselves at Melton in Leicestershire. A pint of elderberries is boiled in a pint of claret (hence the name) and left to stew overnight in the bottom drawer of the Aga. The liquid is drained off and boiled with mace, shallots, peppercorns, cloves and ginger then bottled and stored in the cellar for a year. It is absolutely terrific with cold meat, game, steak or liver and the longer you leave it, the better it be.

Wine has been made from elderberries since time immemorial; it is very easy to make and similar to good burgundy or port, depending on the recipe. My first alcoholic drink, when I was ten years old, was elderberry wine, made by a German ex-prisoner of war called Paul Richter. Paul lived by the ruins of Laughton Castle out on the Laughton Marshes, where my father had the shooting rights. One freezing November evening we went into Paul's cottage after duck flighting and he gave me a glass of the deep purple wine to warm me up. I remember it being the most exquisite thing I had ever tasted. During the eighteenth and nineteenth centuries, acres of elders were cultivated in Kent to make wine, spuriously sold as French clarets and burgundies or as port. A basket of elderberries, water, yeast and sugar is all that is needed for a basic wine and I thoroughly recommend it. Another marvellous drink which knocks sloe gin into a cocked hat, is elderberry schnapps. Simply fill a Kilner jar with elderberries and pour in a bottle of vodka. Stand in sunlight for a day or two, giving it a good shake every so often and leave in a cellar or dark storeroom for a few months. Strain and bottle the schnapps and use the vodka-infused elderberries as a stuffing for wild duck, sauces for game, mousses, sorbets or pies.

Elder has been called nature's 'medicine chest', and the range of medicinal uses for it is quite staggering. Elderflower water, made from boiled water and elder flowers, was once the universal skin softener. They were mixed with clarified lard or goose fat as an early form of face cream and as a soothing lotion for scalds, stings, chilblains and scar tissue. During the First World War, the Blue Cross appealed for elder flowers to make into unguents to treat injured artillery and troop horses. Elderflower teas, possets and syrups were recommended as a blood purifier and in escalating doses for colds, influenza, asthma, bronchitis and pneumonia. The berries, bark and leaves contain viburnic acid and were used to induce perspiration to sweat out fevers, for rheumatism and syphilis, made into an ointment to treat haemorrhoids or mixed with fennel to relieve sciatica.

Autumn Bounty

There is a noticeable change in the season during late August and September, as the nights begin to draw in, temperature drops, valley bottoms fill with mist in the early mornings, swallows start flocking in preparation for their migration home and the summer moorland nesters – curlew, golden plover, snipe and oystercatchers – begin drifting back to the coast. Hedgerows fill with autumn colours as nature now provides us with a cornucopia of nuts and berries: blackberries (*Rubus fructicosus*), which flower from May to September and produce their delicious purple berries from late August until early November. Blackberries can be found all over Britain in bramble bushes along hedgerows, railway embankments, inner-city canal banks, waste ground and woodland fringes. The lowest berry of each cluster is the first to ripen and has the best flavour. Fat and juicy, these were the ones we looked for as children and would endure any number of scratches to get hold of. The secondary berries are picked for preserving, jams, puddings, ice cream, jellies, blackberry wine and for flavouring vinegar. Traditionally, blackberries were over by the end of the first week in October and an old country saying, 'The Devil pisses on blackberries in October,' was a reference to the surviving berries turning soft with the first frosts of winter. Nowadays, with milder winters, the smaller berries can still be picked for several weeks more, although they tend to have a higher proportion of pips and are best used in pies with other fruit.

Not so long ago, blackberry picking in September was a significant event in the lives of urban people. It was one of the ways in which they kept in touch with their rural background, and every weekend whole families, armed with baskets and buckets, would descend on the countryside by bicycle, train, motorcycle and sidecar or buses especially booked for the occasion. Sadly, the custom died out in the 1960s and 1970s, as hedgerows that some families had visited every year were bulldozed out to create large field units. Preserving blackberries for winter consumption was a valuable source of vitamin C, and the leaves, bark and root, which contain a high proportion of tannin, were made into a decoction during the sixteenth and seventeenth centuries for treating dysentery.

Wild roses (*Rosa canina*) or dog rose (from dag or dagger – a reference to the wickedly sharp thorns) are common on waste ground, woodland margins and hedgerows. It is one of the traditional hedging plants and millions were planted with hawthorn, holly, hazel, beech and blackthorn during the Acts of Enclosures, when 350,000 km of hedging was established between 1603 and 1850. The rambling tendrils of wild roses produce fine pink or white blossom from June to July. Towards the end of the flowering period, the petals become loose and can be plucked off for use in salads or as a flavouring for sweet water, vinegar, jellies and sorbets. They can be made into wine, jam, crystallised or used to add a delicate fragrance to puddings – they are particularly good with rhubarb. Rose hips, the bright red fruit which appear from late August until November, contain twenty times as much vitamin C as oranges. In medieval times they were boiled into a purée with honey as a pudding, but rose hips have been principally used for the last century in the form of a syrup as a vitamin-C supplement through the winter. Rose hips can also be made into wine or a rich soup using fresh or dried hips.

Blackthorn (*Prunus spinosa*), the hardy thorn bush found all over Britain on heaths and in hedgerows, produces bitter blue-black sloes in September and October. Traditionally, sloes were never picked before the first frosts, which softened the fruit and removed some of the unpalatable acidity. Now the same thing can be achieved by putting them in the deep freeze for a couple of days. Most sloes are used to make the ever-popular sloe gin, a liqueur easily made by filling a 2-litre Kilner jar half-full of frozen sloes that have been pricked with a fork. Add 50 grams of crushed barley sugar and a bottle of gin. Leave in a warm place for at least three months, giving it a good shake from time to time. Strain through muslin into bottles, but keep the gin-infused sloes for stuffing duck or for a rich sauce to accompany any game. Sloes also make a superb wine similar to port, and in the eighteenth century, litres of sloe wine were sold by fraudulent wine merchants as genuine port. Sloe jelly is vastly superior to redcurrant with lamb or venison.

Another marvellous fruit of hedgerows, scrub woodland and heaths are little, rock-hard, yellow crab apples (*Malus sylvestris*). Wild apple trees, with their gnarled twisted limbs, produce a beautifully scented flower in May and fruit from August to November, which are at their best from early September until the middle of October. Crab apples were highly prized by ancient Britons both as a food and as an ingredient in a very potent, cider-based mead made from fermented apples and honey. All manner of myths and folklore became attached to crab apples, the most abiding of which was the Twelfth Night ceremony of Wassail. Originally part of a pagan fertility rite intended to hasten spring, wassail involved drinking toasts to fruit-bearing trees. It later became part of the Christian ritual of goodwill to mankind with honey sweetened, hot spiced ale, in which roasted crab apples floated, drunk from beautifully carved wassail 'loving cups'. Although the wassailing revels have died out in most parts of the country, orchard-visiting wassails continue in the West Country with Whimple in Devon and Carshalton in Somerset holding particularly famous ones on Old Twelfth Night, 17 January. On this night, villagers caper round the largest apple tree in a local orchard, drinking mulled cider and yodelling the wassail song:

> *Old apple tree, old apple tree;*
> *We've come to wassail thee;*
> *To bear and to bow apples enow;*
> *Hats full, caps full, three-bushel bags full,*
> *Barn floors full and a little heap under the stairs.*

Shotguns are fired to frighten away evil spirits and pieces of cider-soaked toast are hung from the branches for robins, which are considered to represent the good spirits of the tree. Crab apples are too bitter to eat raw – the name comes from the old English word meaning sour, bitter or twisted, and in Scotland the word 'crabbit' is often used to describe an evil-tempered woman. One of the strangest customs in Britain is the Egremont Crab Fair, which has been held every September almost continually since the Cumbrian town was granted a Royal Charter in 1267 by Henry III. A principal feature of the fair is the World Gurning Championships, where contestants compete to pull the ugliest face whilst their head is stuck through a horse collar. This extraordinary practice, where competitors have been known to devote a lifetime to achieving exceptional ugliness, was inadvertently started by Thomas de Multon, Lord of the Barony of Egremont. After harvest, de Multon was in the generous habit of rewarding his serfs by riding through the town and tossing each a crab apple; the bitter taste of the apples caused the peasants' faces to contort and thus began the tradition. Giving away crab apples continues to this day with the 'Parade of the Apple Cart', where apples are thrown to the people who line the main street.

Crab apples may be inedible raw, but are very good in their cooked form, especially when mixed with a sweeter fruit, such as blackberries. Crab apple jelly has an exquisite flavour, which is particularly delectable with hot scones or toast. Crab apple cheese, made with cloves, cinnamon and nutmeg is delicious as a pudding or served as a sauce with goose, pork or boiled ham. Crab apple wine, a medium-sweet modern version of the brew drunk by the ancient Britons, is easy to make. We pickle crab apples in cider vinegar and Barbados sugar every autumn, to eat with our Bradenham ham at Christmas — their tart flavour offsets the fattiness of the ham. We also make verjuice by roughly mashing them in a food processor, letting the mash ferment with a little water for four days; straining through muslin and bottling. It is a terrific alternative in all dishes where vinegar or lemon juice might otherwise have been used.

Rowan or mountain ash (*Rorbus aucuparia*), whose profusion of pale pink, early summer flowers and clusters of brilliant orange autumn berries has made it a favourite with municipal landscape planners, is common on dry heaths and woodland, moorland fringes and rocky valleys. There is much superstition and folklore connected to Rowan trees, particularly as a ward against witchcraft. Rowans were one of the Celtic sacred trees and the berries were used by them to flavour mead or ale. I have known several hill shepherds who made wine from Rowan berries and a retired naval commander who used the juice instead of Angostura bitters to make pink gin. For many centuries, rowan berries, sometimes mixed with crab apples, have been used to make a rather sharp-tasting jelly, which is the traditional accompaniment for venison or roast mutton. They make an excellent stuffing, fresh or dried, for old grouse or wild duck.

Hazel (*Corylus avellana*) was one of the traditional hedging plants and can be found in old hedgerows, scrubland and as understorey woods across Britain. Hazels produce male 'lamb's tail' catkins and buds with little red protruding flowers from January to April. The nuts, encased in their thick frilled green husk – the name is from the Anglo-Saxon word, *haesel*, meaning cap or hood – appear from late August to early November. Hazelnuts are high in protein and carbohydrates, and nutting for hazelnuts was another important custom involving the whole family which has died out. Up until the 1930s, it was common practice for the schools of villages and market towns to close on Holy Rood Day, 14 September, to allow children to spend the day gathering nuts with their parents. Once picked, the nuts must be kept in their shells in a warm dry place for eating through the winter, simply as a nut or used in an infinite variety of pudding dishes: tortes, macaroons, pies, meringue, ice cream biscuits or as paste fillings for cakes.

Sweet chestnuts (*Castanea sativa*), described as the most magnificent trees in Europe, are widespread throughout England, particularly where such a tree would enhance the immediate locality, such as parks, Royal forests, near villages or in woods. They produce flowers in June and July which take the form of upright catkins, with the male flower on the upper part and the female on the lower. The female parts develop into fruit-bearing spiny cupules in the autumn, which ripen and fall with the leaves in October and November. I love hunting for prickly chestnut husks among the golden leaf carpet of a wood in the chill of a late October afternoon, with the musty smell of damp earth, an evening mist beginning to creep into the hollows and every so often the skittering sound of an animal hurrying through the leaves. Sweet chestnuts are tremendously versatile and were considered a delicacy by the Romans; it is generally assumed that it was they who introduced the tree to Britain. The nuts can be pickled, cooked with cabbage or Brussels sprouts and made into soup; puréed or used as a stuffing for pheasant, chicken, and, of course, turkey. If there are any really big ones, they can be lightly boiled and repeatedly dipped in hot syrup to make my favourite Christmas treat – marron glacé. Or they can be simply roasted in the embers of a hot fire as part of the tradition of Christmas. John Evelyn succinctly described sweet chestnuts as, 'delicacies for princes and a lusty and masculine food for rusticks, and able to make women well-complexioned'.

Wild Fungi

There is a sense of urgency among wildlife in autumn, as the leaves begin to turn, daylight hours become shorter and the food source for many of them diminishes. After a summer of sybaritic ease, drone bees are brutally evicted from their hives with biblical ferocity and creatures that go into partial hibernation – bats, toads, snakes, hedgehogs and the smaller mammals – are feeding hard to put on body fat for winter. The intake now, for those that have had second litters or were late born, will be the difference between life and death. Jays, grey squirrels and our few remaining reds start to hoard acorns and beech nuts; mushrooms and toadstools start to emerge and, on warm evenings, the last scents of summer flowers are mixed with a faint whiff of decay. Nothing is more synonymous with this period of mists and mellow fruitfulness than the mushrooms, toadstools and other fungi that appear in our woods and old pastures.

Despite the vibrant food culture of the last couple of decades, the prevalence of cultivated fungi in supermarkets, dried fungi in every aspiring foodie's larder, or the educational fungus-identifying forays organised by conservation groups, we as a nation – except for dedicated aficionados, who really know what to look for – have a deep-rooted antipathy to picking fungi in the wild. Why this should be, when all our Continental neighbours have an historical culture of harvesting edible fungi, I have never quite understood. None of the explanations – that the early Christian Church anathematised fungi because hallucinogenic liberty cap and fly agaric had been part of Druidical ceremonies, or that in Roman Catholic countries fungi were acceptable Friday fare – carry much substance. I think the answer lies in a thoroughly British suspicion of anything unusual in the food line, mixed with inherent, basic common sense.

Requiring no light, most fungi emerge in dank, dark, sinister places where no healthy plant would be expected to survive, many of which are fantastical organisms that sprout up among leaf litter, decaying stumps, the trunks of trees or simply emerge as jelly, oozing out of the ground. These prehistoric parasitic growths, lacking the ability to produce their own food, absorb sustenance from living or dead plant matter, creating a nutrient recycling process without which crops, trees and grass would not survive. There are something in the region of one and a half million different species of fungi worldwide, which include moulds, rusts, smuts, yeasts, mildews, lichens, mushrooms and toadstools, with perhaps as many as 20,000 in Britain.

Many, like Jew's ear, yellow brain and tripe fungus resemble discarded body parts, so succinctly described by Shelley as 'pale, fleshy, as if the decaying dead with a spirit of growth had been animated.' Some, stinkhorn and witch's egg, have the sickly sweet smell of decomposing carrion. Others, known to be edible, bore too close a resemblance to those that were deadly poisonous: delicious little chanterelles, which appear in carpets of eggshell yellow among mosses in deciduous woodland, resemble the aptly named deadly webcap. Pale-skinned common morel could be mistaken for the toxic false morel, with fatal consequences to the liver and kidneys. Confusing common field and horse mushrooms with death caps or destroying angels was far too big a risk to take. Most boletus, which appear very similar, could be eaten with safety, but by no means all. Shaggy inkcap, which grows on lawns, roadsides and garden rubbish heaps, has an almost identical nasty cousin: common inkcap – the medieval source of ink – which contains the toxin coprine. Coprine becomes activated if the fungus is ingested with alcohol and the effect, although not necessarily terminal, is enough to put someone off any sort of mushroom for life. In the sixteenth century, John Gerard, an apothecary to James I and author of the famous *Herball*, summed up the prevailing view of fungi with the warning that, 'most of them do strangle and suffocate the eater.'

None of these inhibitions affect wildlife and fungi play an essential role in the food chain. Fungi provide specialist habitats for a huge number of mini beasts and their larvae, which in turn assist the fungi by distributing their spores. They are a host for flies such as fungus gnats, phorids, thrips and minute cecids, bluebottles, green bottles and flesh flies; any number of beetles, including dor beetles that hollow out bracket fungi, ambrosia beetles and the elm bark beetle that carry the spores of fungi from tree to tree. The great black, ash-grey and tree slugs are all voracious fungi eaters, and so too, are rabbits, badgers, hedgehogs, deer and red squirrels. Both deer and hedgehogs have fungi named after them: the little chunky hedgehog fungus, much prized by European epicures; *Pluteus cervinus*, an umber-coloured mushroom of deciduous woodland, much favoured by fallow deer; and the descriptively named deer balls – a globular yellowish toadstool, once used medicinally to encourage lactation. Our few remaining red squirrels supplement their winter supplies with a variety of the boletus species that grow in coniferous forests, carrying them up into the over mantle and fastening them to twigs or among conifer needles.

If fear of poisoning led to fungi being shunned as food, there were many other practical uses for them. Birch polypore or razor-strop fungus, the rubbery white bracket fungus which grows at right angles on birch trees, was harvested in large numbers and sold to barbers and wood carvers to put a fine edge on cutting tools. Birch polypore contains antiseptic properties and was used by Highlanders to pad their targes (shields) against impact and, if necessary, as wound dressings. Another birch polypore, the hoof or tinder fungus was, as the name suggests, an essential component of a tinderbox. Later on, it was used in the manufacture of 'fusees' an early form of match sold under the name of Amadou. The hard, corky fungi were beaten flat, cut into thin strips and the ends dipped in saltpetre. Hoof fungi were also used medicinally as a styptic to control haemorrhages. Green wood-cup colonises the fallen branches of oak trees, staining the wood a beautiful verdigris colour. Spores of this fungus were deliberately introduced to timber planking by eighteenth-century cabinet makers to create a curious colouring known as 'green oak' used in Tunbridge ware, a unique form of marquetry. The wood of oak trees infected by ox tongue fungi was found to have a darker colour and was much sought after by furniture makers. Jew's ear, the slimy brown growths often seen on elder trees, soothed and healed ophthalmic complaints. Powder from dried giant puffballs had a variety of applications: it made good tinder; beekeepers used it — and still do — when smoking bees; surgeons treated septic wounds with it to draw foreign matter and it could be made into an invigorating tonic. Fly agaric, the highly toxic scarlet mushroom found in birch woods, was pulped, mixed with water and painted round doors and windows to kill flies.

Curious stuff, fly agaric. A powerful hallucinogenic, it was used for centuries as an inebriant by the nomadic reindeer-herding tribes of northern Scandinavia and Russia. The habit started, apparently, when the tribesmen noticed the effect of the hallucinogen on reindeer, who find the mushroom palatable. The popular method of ingesting the drug was to tether the inebriated beast and drink its urine. Muscimol, the psychoactive element remains active in urine and revelling tribespeople were able to recycle the drug among themselves, for as many as seven re-ingestions. Maddeningly, history fails to tell us how this fascinatingly economical way of holding a party was ever discovered but it is indubitably the origin of the expression 'to get pissed'.

For those who are expert enough to distinguish between edible species and their poisonous lookalikes, there are some wonderful delicacies to be found in the autumn. But, having known someone who made a simple mistake and spent the rest of her life on a kidney dialysis machine, I cannot emphasise the importance of studying fungi with a professional mycologist before contemplating foraging. Following the resurgence of interest in wild food, there is now no shortage of expertise available through fungi identifying courses and workshops run by local fungus recording groups, the British Mycology Society, the Field Studies Council or natural history associations. A number of cookery schools and fine-food outlets also run courses and some of the most enjoyable are those offered by Valvona & Crolla, the famous Edinburgh delicatessen. During September and October, Valvona & Crolla organise tutored fungi-identifying expeditions led by a consultant mycologist. Parties are taken by bus to woodland locations likely to provide a wide variety of fungi, spending a happy morning among the trees, collecting any fungus they come across. They all meet up again at lunchtime for a terrific picnic and selection of Italian wines provided by the delicatessen, whilst the mycologist identifies and discusses their findings.

The most delicious fungi are often among the easiest to find. One of my all-time favourites is giant puffball (*Langermannia gigantea*). They are not uncommon in woodland glades, the corners of fields and under hedgerows, emerging from August to October. Puffballs are usually between 10 and 30 cm in size, but can sometimes be found much larger and it is a real bonanza to stumble across one of these gross white globules glinting in the autumn sunlight. One of their advantages is being unmistakable for anything else; as long as it is pure white and more or less spherical, it's a puffball. If the skin has a yellowish tinge, they are past their best and will soon turn brown, exuding a cloud of dust spores when kicked. The best way to eat one is to cut 1-cm slices from top to bottom and fry them in bacon fat until golden-brown. Pink-gilled field mushrooms (*Agaricus campestris*) and the larger horse mushrooms (*Agaricus arvensis*) are usually easy to find from August to November in pastures and old meadows, particularly where horses are kept. The wonderful flavour of horse and field mushrooms is best appreciated if simply simmered in a little butter or milk. Unfortunately, yellow-staining mushroom is almost identical, except for the stem, which turns bright yellow when cut and can cause acute digestive upset.

The scaly brown parasol mushroom (*Lepiota procera*) is unmistakable in wood margins, on stubble edges and along roadsides. It rises, closed, like an old-fashioned sun shade, the top held by a loose ring of tissue which moves up the slender stem as it opens. This eventually breaks free and can be moved up and down the stem. Parasol mushrooms are about 18 cm when fully open and are good cooked as field mushrooms or sliced, dipped in batter and fried. Another edible species, shaggy parasol (*Lepiota rhacodes*) is similar, but with a scalier cap. Ceps (*Boletus edulis*) are common in the open spaces of deciduous woodland, particularly beech. They have rounded, shiny brown caps supported by a spongy mass of pores, which represent the gills found in other mushrooms, and short, stubby, stems. Their country name is 'penny bun,' which they resemble to some extent and a sweep of ceps, growing among fallen beech leaves is a joy to find. Ceps have a wonderful nutty, earthy taste – '*L'eau de terre*' – and are one of the most popular fungi on the Continent. There are a prodigious number of recipes for them: they can be grilled, baked, sautéed, stewed and fried; added to casseroles or soups; dried for winter consumption and made into flour. Boletus come into a number of species, all of which are very good to eat, except *boletus satanas*. Satan's boletus looks very similar to all the others; it is easily recognisable to those who know what they are doing but is potentially, an extremely unpleasant mistake for those who don't.

Chanterelle or girolle (*Cantharellus cibarius*) are beautiful little trumpet-shaped fungus of pine, oak, birch and beech woodland and appear in golden carpets from July until the first winter frosts kill them off. They are the most sought-after fungi on the Continent and have an exquisite flavour – a handful of freshly picked chanterelles has the aroma of fresh apricots. They are superb sautéed in butter with diced potatoes and bacon; fried with a little oil and garlic or in an omelette. Deadly webcap have been mistaken for chanterelle, with devastating results; one positive way of telling the difference is that the gills of chanterelle run down the stem. However, always remember the maxim of even the most experienced fungus forager: 'If in doubt, leave well alone.' Oyster mushrooms (*Pleurotus ostreatus*) sprout in grey clusters from the dead or dying branches of beech or ash trees during the autumn. As with many fungi that grow on trees, they tend to be rather tough, but young specimens, sliced, dipped in egg and breadcrumbs and deep fried, are very good. They can also be added to stews, casserole or soups, or grilled or dried for winter consumption. Chicken of the woods (*Laetiporus sulphureus*) can be found at any time of the year, except late winter, in gibbous yellow layers on old oaks, yew, sweet chestnut or willow. This fungus should be cut whilst it still has the yellow vibrancy of youth. This is a meaty fungus and when chopped into cubes, their texture and rich flavour are a delicious addition to casseroles.

Shaggy inkcaps (*Coprinus comatus*) or lawyer's wig are often found clusters in fields, road verges, canal banks, garden rubbish tips – virtually anywhere that humans have disturbed the ground. They emerge like white guardsmen's busbies, covered in light-brown scales. As the slender stem grows, the fungus partially opens like an umbrella, with pink gills that turn black and liquid as the fungus matures. Eventually it dissolves into an inky goo. Picked young, when the gills are still pink and eaten almost immediately, they are delicious sautéed in butter or baked with eggs and a little cream. I find the flavour rather elusive and prefer to make them into ketchup by packing layers of the fungi with salt in an earthenware crock. The resulting fluid is strained, seasoned with pepper and nutmeg; brought to the boil and bottled. Shaggy inkcap is similar to common inkcap (*Coprinus atrementarius*), a mushroom which reacts badly with alcohol. It is distinguishable by being more slender than shaggy inkcap, a lack of scales and a dingy grey colour.

Searching for nature's bounty and being rewarded with some delicacy, fragrance or even, if one knows what one is looking for, a palliative, is a cathartic experience which gives the forager a greater understanding of the land and the changing seasons. The wild places where these treasures grow, must be preserved and cherished.

Capercaillie

For many of my contemporaries in the late 1960s and 1970s, an invitation to stay with Harry Calvert and his parents, Major Eddy and Mrs Calvert at Fasnakyle, their 8,095-hectare deer forest in the east end of Glen Affric, was a first introduction to the glories of a Highland estate. There were ten pools on the River Glass, stags in the rut, walked up grouse, ptarmigan of the high ground and reputedly, a few capercaillie in the old Scots pine woodland behind the lodge. The Calverts were immensely kind and generous hosts, and between them, Major Eddy and his stalker, Ian Shewan, went to endless trouble to share their great depth of knowledge with the young.

As a Borderer, I had never seen a live caper and one afternoon, when we were down early off the hill and the rest of the party had gone to fish the Glass, Ian offered to take Harry and I up into the pinewoods and try to drive a cock caper he had been watching for some days over us. Even after the passage of fifty years, I vividly remember climbing up through the old self-seeded trees, with shafts of early October afternoon sunlight shining through the high canopy, until we came to a wide, sunlit open glade, where blaeberry grew among the rocks and moss-covered fallen trunks. This was where Ian had seen the cock feeding in the early mornings and popping in a couple of No. 3s, we got into position on either side of the glade and waited in hopeful anticipation. The wood was utterly silent until we heard Ian tapping towards us and then his whistle, indicating the caper was on the wing. What happened next, was so completely the opposite of what I expected, it has remained imprinted on my mind; a great, black, elongated shape, slid noiselessly at surprising height and speed through the trees on my right and disappeared from view. I was so astonished, I did not even put my gun up.

By comparison, a hen caper is a dowdy little thing, similar to and not much bigger than a grey hen, whereas the cock is a magnificent, great brute of a bird standing a metre high and weighing as much as 8 kg, with iridescent black plumage, scarlet head wattles and the powerful curved beak of an eagle. On the ground he resembles something of a turkey cock, and I had expected ours to behave much the same way in flight, but those who have shot driven caper will tell you that for all his size, he is a fast, dexterous, hard-flying bird and extremely difficult to bring down.

The cold wet springs and summers of the Little Ice Age in the late seventeenth and early eighteenth centuries, and extensive felling of the capercaillies' open pinewood habitat, with its isolated mature trees and horizontal branches for roosting and winter feeding, and a floor of the blaeberries which make up their summer diet, led to them becoming extinct by the mid-eighteenth century. In 1836, Sir Thomas Fowlett Buxton of Northrepps Hall, Norfolk, arranged for thirteen cock caper and sixteen hens to be sent from Sweden to Taymouth Castle as gift for his friend, the Marquis of Breadalbane. These were followed by two further consignments and in 1839, James Guthrie, Lord Breadalbane's keeper, reported that seventy caper were successfully established and breeding in the woods around the castle. The large-scale replanting of the previous century on Scottish estates such as Atholl, Scone, Breadalbane, Blair Drummond, Invercauld, Monymusk, Mar Lodge and Moray – the Earl of Moray is reputed to have planted over 10 million Scots pine trees from seedlings collected on his estate – and many others north of the River Tay, must have provided highly attractive habitat, as the caper population expanded at an extraordinary rate. By the turn of the century, caper had advanced from Taymouth north-east and south-west along the river valleys and were breeding in Argyllshire, Stirlingshire, Aberdeenshire and as far north as Golspie in Sutherland.

The first reintroduced caper were shot at Taymouth in September 1842, when Queen Victoria and Prince Albert were staying with Lord Breadalbane. Among the large shooting party which included Prince Albert and the Marquis, was Sir Thomas Fowlett Buxton, who recorded that the bag for the day included a brace of 'very wild' capercaillie. As the population spread and expanded, caper appear regularly in bag records of estates where there was suitable habit – at Atholl for example, His Grace's game book lists 337 caper cocks and hens shot between 1865 and 1916, and it is reasonable to assume that these would have been part of the bag on driven days.

Very little has been written about shooting driven capercaillie and first-hand accounts are even harder to find, as those who remember them when they formed part of a day's shooting are now few and far between. In *Shooting Days* (1918), Eric Parker, editor of *The Field* from 1931 to 1937, gives a vivid description of caper coming through during a driven pheasant day in Angus just before the First World War. On a bitter January afternoon smelling of snow, guns were lined out for the last drive on a steep slope amid rocks and bracken, facing a high pinewood. The drive started with fast, agile wild pheasant followed by a crash and rattle in the pine tops, and warning shouts from far back alerting them that caper were on their way: 'Nothing more, 'til they swept through the tree tops, first a hen, then cocks and hens in twos and threes. They came very silently, swinging out from the hillside across the valley; the vision remains with me of a grey and yellow sky, dark pines and great birds floating out on level wings into the sunset'.

J. K. Stanford, the sporting journalist, author and ornithologist, mentions trying to shoot caper between the wars in *Wandering Gun* (1960) and then devotes several pages to a fascinating account of a driven caper shoot in November 1953, as a guest of Sir Alan McLean of Littlewood Estate, Donside, Aberdeenshire. Over the five drives, caper seemed to have come from every height and angle – on one drive, caper flushed far back in the wood, actually flew in a wide circuit and came in over guns from behind – but all were completely silent and incredibly fast, even when driven against the wind. 'All day,' Stanford wrote, 'I seemed to have about one second between a caper appearing in front and vanishing behind' and the few that were hit on the snout always crashed down far behind the guns. Stanford, who had shot every sort of driven game in Britain, India and Burma, observed that he had never experienced the same degree of haste and lack of time as he had with caper and that 'their silence, their speed and the momentary target they offered, were to me more impressive than their size when picked up'. The bag for the day was thirteen caper and in a good year, thirty to forty were shot at Littlewood.

Mick Benson, the head keeper on the Tanfield Estate in North Yorkshire, remembers his father, Brian, describing a caper-only shoot in Morayshire in the mid-1970s and the details of the day in his gamebook. Brian was keeper on the Army Shoot at Ripon for forty years and was invited by one of the officers who had an estate near Tomintoul where there was a shootable surplus. Guns were placed along the bottom of a ravine with old pinewoods on either side and as with others who have shot driven caper, Brian was astounded at their speed, silence and how entirely different they looked on the wing; 'Not unlike a cormorant in the way their necks stretched out and bodies became elongated.' When the drive started, hens seemed to come through first, with cocks following the same line. Guns were limited to two cocks each and on the second drive, after an incredible left and right, Brian decided to spend the rest of the day with the beaters. Capers are extremely wild, always highly alert and of course, widely dispersed in the woods; funnelling them towards a line of guns and preventing them flying out or back was a delicate operation. Brian remembers that absolute silence was required from the beaters as they got into position to start a drive and that no dogs were allowed.

Another who shot driven caper back in the 1970s, is Barry Aston from Leeds, a keen wildfowler who had gone up with a few friends to flight geese on Loch Eye, near Tain. During the course of their stay, they met some keepers in a local pub, who mentioned they would be reducing the caper population on a neighbouring estate and invited them for the day. Even to people used to shooting geese, Barry and his friends found the caper extremely challenging, but the thing that sticks in his mind, is winging a cock caper just as the beaters came through on the last drive and turning round to find it running towards him. Unaware of the cock caper's reputation for aggression, he was surprised to hear the beaters shouting to keep his dog back. His other memory is having one of the capers cooked back at the lodge they had rented and trying to eat it; 'It was revolting. Tasted of turpentine. Even the gravy was revolting.'

This reinforces the popular view of caper as a table bird, but Peter Fraser, the head stalker at Invercauld for forty-three years until 2012 and current vice chairman of the Scottish Gamekeepers Association, disagrees. A caper shot in the winter months when it had been feeding on pine needles would undoubtedly taste resinous, but one shot in early October (the season was 30 September to 1 February) after a summer on blaeberries, or those he remembered as a child growing up near Ballater, when they were sometimes shot coming out of the woods to feed on gleanings after the oat harvest, were delicious; pale fleshed and less gamey than a grouse. I asked Peter whether caper were shot at Invercauld. 'Only occasionally,' he told me, when a guest wanted to shoot one and the keepers would try to oblige, but he did remember organised caper culls on neighbouring estates and Forestry Commission land to protect new plantings. The massive post-Second

World War commercial replanting schemes of species such as Sitka spruce, appeared to have temporarily benefited the caper population and by 1970, it had reached 20,000. Considerable damage was caused by caper eating the palatable buds and shoots of seedlings, or breaking the branches of saplings as they landed on them, particularly in the lekking areas.

The population collapsed in the 1980s to the extent that estate owners introduced a voluntary moratorium on shooting caper in 1990 and by the time they became a protected species in 2001, there were fewer than 2,000 left. A number of factors have contributed to the decline: habitat loss – the fate of any capercaillie population is tied to the quality and quantity of its habitat; the dark, serried ranks of commercial conifers are completely alien to them. Another cycle of cold wet springs with hens in poor condition when they go to lay and the inevitable reduction in brood survival rates, plus the ever-increasing chick mortality from predators. Capers are ground nesters and the chicks are vulnerable to crows, foxes, raptors, badgers and pine martens, the last three are themselves protected by law. Human disturbance at lekking and nesting plays a part, and in all probability, since they affect other woodland and moorland creatures, the blood-sucking parasite tick. The capercaillies' dependence on and their exacting needs of a diverse but specific habitat, led to their extinction in the past and despite conservation efforts by landlords, the Forestry Commission and RSPB, the population of these wonderful game birds has now tragically, dwindled to little more than a thousand.

Winter

Ferrets

Now that the closed season for inland game birds is upon us, I sometimes break the ennui by popping my polecat hob into his travelling box on crisp February mornings and spending an hour or two having a go at bolting bunnies to my old Holland & Holland single-barrelled .410. In many ways, these little rabbiting forays are a trip down memory lane; I was first taken ferreting by our old gardener Joe Botting, when I was six years old and like generations of other small boys, ferreting was my first introduction to fieldcraft and the responsibilities that go with looking after a working animal. More importantly, after a little initial tuition, bolting rabbits to nets was something I could do without adult supervision and learn from my own mistakes. Success depended on silence, identifying wind direction, proper care of the equipment and sensible treatment of a creature that bites hard if handled stupidly. Ferrets are also wonderfully versatile – over the years I have had enormous fun ferreting rabbits to nets, Harris hawks or whippets and there is no finer training in handling a gun. To shoot bolting rabbits, a novice must be quick getting his gun up and shoot fast. The mistake common to all shooting – being behind or below – is quickly identified by the telltale puff of dust on the heels of a departing bunny.

A true ferret, the descendant of either a Steppe polecat, a European polecat or a hybridisation of both, is yellowish in colour and has the red eyes of a genuine albino. Quite when and where the first polecat was domesticated has been lost in the mists of antiquity, but their sporting lineage must be as old as hawk or hound. The nomadic Black Sea tribes used nicotine-coloured Steppe polecats to bolt their staple diet of sousliks and marmots at least 3,000 years ago. Ferrets were well known to the ancient Greeks and were used by the Romans to hunt rabbits in Spain and the Balearic Islands. Although the tradition continued in Europe, the exact date of the introduction of rabbits and by association ferrets to Britain, is uncertain. Rabbits were among the species in the living larders imported by the Romans – rabbit embryos, laurices, were considered great delicacies – but none appear to have survived the collapse of the Empire. There is apparently no native name for rabbit in the Anglo-Saxon language, nor any mention of them in the Domesday Book. Most historians agree that they were brought across from Europe around the middle of the twelfth century as part of the existing and highly lucrative culture of rabbit farming.

Warrens or coney garths, with their pillow mounds to encourage burrowing, became established all over Britain wherever there was suitable free-draining soil. The rights to establish a warren were franchised or gifted by the king and some were huge places such as Lakenheath in the Brecklands of Suffolk, set up by the Bishops of Ely, surrounded by a bank 10 miles long topped with furze and patrolled by armed guards. Rabbit farming was a very valuable part of the medieval economy – the flesh was an important source of winter sustenance and the skins, worth proportionately as much as a bullock's. Inevitably, some rabbits escaped and established themselves in the surrounding countryside. At some stage, peasants observing the warreners at work, will have caught and tamed wild polecats and ferreting became the poacher's bread and butter. A risky business – the Game Laws of 1390 imposed a year's imprisonment on anyone possessing ferrets who did not have 'in lands 40 shillings per annum'. The rabbit population rocketed during the agricultural improvements of the eighteenth and nineteenth centuries and ferreting became a necessary form of vermin control. At the same time, with the shift in population from country to town during the Industrial Revolution, ferret keeping was for many urban people, a link to their rural background.

For most of their long history, ferrets tended to be poorly treated and most people, such as Joe, kept them in hutches which were far too short for active animals, feeding them on bread and milk, believing this made them easier to handle. Curiously, the introduction of myxomatosis in the 1950s was the catalyst which eventually led to enormous advances in ferret welfare. The majority of owners hung on to their ferrets regardless of the shortage of rabbits, to discover that like all Mustelidae, ferrets are highly intelligent natural comics which make delightful pets. Improvements in housing, diet and veterinary care were all driven by an increase in pet ferret ownership now so large, that ferrets were voted the pet of the Millennium. There are over fifty ferret rescue centres devoted to the welfare and rehabilitation of ferrets which have become lost or been abandoned by unscrupulous owners. Many of these organise enormously popular ferret racing championships – there is an annual entry for the fastest in the *Guinness World Records* – to raise funds and promote the image of ferrets, all of which was to benefit working ferrets as the rabbit population recovered and the interest in ferreting returned.

In my day, a sporting child had to rely on someone like Joe to get them started. Nowadays, there are hundreds of ferret clubs eager for new members and with the emphasis on educating the young of all backgrounds about their rural heritage, there is no shortage of experts in BASC or the Countryside Alliance to offer help and advice.

Otter

The wheel of the seasons changes with the winter solstice, the longest night and shortest day between 21 and 22 December, when the sun at noon reaches its lowest altitude above the horizon. The end of December is a strange time of year; nature seems to vibrate with the anticipation of lengthening daylight, yet it is really only the start of winter – temperatures are at their worst and the battle for survival among wildlife at its most critical.

Last winter, there was high pressure just before Christmas, with a bitter north-easterly howling down from Iceland. Temperatures plummeted to minus 18, pipes burst, ice formed on the inside of windows and the bristles of my toothbrush froze. The Hermitage and Liddle Waters on either side of the farm became sheets of ice, hunting stopped and the Borders, from Carlisle to Berwick-on-Tweed, wore a silver pelt of hoar frost. The wind had dropped when I took the terriers for their nightly run on Christmas Eve. The ground crackled with every footstep and the gaunt bastions of Hermitage Castle sparkled in the beam of the Long Night Moon. A vixen screamed her eerie, primeval mating call up on the Castle Brae and was answered by the distant triple bark of a dog fox, somewhere on the hillside across the valley. How appropriate, I thought, that on this particular night these two should be celebrating the start of a new season.

At that point, my reverie was broken by the terriers racing back across a field by the Hermitage Water, their muzzles glued to the ground. Up over the bank they sped and in through a window of an old cart shed in the steading, where they started the ferocious clamour only terriers are capable of when working each other up to attack something bigger and stronger. Rushing in with the torch I was astonished to see a young dog otter, fangs bared and head weaving from side to side, rearing back and hissing like a cat in the corner of the shed. It is indicative of how cold and hungry the poor boy must have been that he was forced to seek shelter so close to human habitation. With great difficulty, I dragged the terriers away and left him in peace.

Although otters will eat frogs and waterfowl chicks in the spring, fish of all sorts are their principal diet and for many centuries they were considered predatory vermin, particularly during the long period when the fish of our inland waters were a vital food source. Otters are crepuscular, lying up in their holt under a bank or hollow tree beside a river and issuing forth to hunt at night. The only effective control was with hounds, following their drag or scent to the holt and bolting them with terriers, the ancestors of the Sealyham, Skye, Cairn, Dandie Dinmont and Welsh. In *Otter Hunting*, by the Earl of Coventry and Captain L. C. R. Cameron (1938), the authors assert that the Celts, who were great hound breeders, hunted otters with rough-coated otterhounds. After the Norman Conquest, every principal building, be it palace, castle, manor or monastery, had stew ponds stocked with carp and inland fisheries were run on a highly efficient commercial basis.

Protecting fish stocks became increasingly important and the first record of an official otterhound pack is in 1175, when Henry II established a pack of Griffon de Bresse hounds and appointed Roger Follo as the King's Otterer, a position that came with a large manor, known as Otterer's Fee, near Aylesbury in Buckinghamshire. All successive monarchs appointed Masters of Royal Otter Hounds and maintained packs for the public good, with Charles II being the last to keep a Royal kennel.

Otters continued to be selectively controlled by hounds, first by private packs and later by the twenty-three packs registered with the Association of Masters of Otterhounds, until 1977, when the association passed a voluntary moratorium on hunting otters. By then the once numerous population had been decimated in England and Wales through river pollution, loss of habitat and agricultural pesticides. In 1978, otters became an endangered species protected by law.

In the past fifteen years, due to the conservation work of organisations such as the Otter Trust, Vincent Wildlife Trust, Environment Agency and the Otters and Rivers project run by Wildlife Trusts, otters have made a remarkable recovery. In some areas they are back to pre-war numbers. Otters have become established on waterways in thirteen major cities, including London, Bristol, Manchester and Birmingham, with sightings of them in more than 100 other urban settings.

Depending on their physical condition, otters can breed at any time of year, although most cubs are born in May and June. Cubs are taught to swim and hunt by their mother at about sixteen weeks and should be independent a month later. A late-born cub is unlikely to survive a very hard winter and the one in the shed would fall into that category. Throughout that week our unexpected Christmas guest lived around the steading, eating mackerel from a summer fishing expedition, placed where I knew he would find them. When the thaw came he disappeared, but I hope the spraints and marks of an otter's rudder I sometimes see in the silt along the river belong to him.

Pike

No warmth, no cheerfulness, no healthful ease,
No comfortable feel in any member —
No shade, no shine, no butterflies, no bees,
No fruits, no flowers, no leaves, no birds —
November!

                                    Thomas Hood, 'November', 1844

Without doubt this is my favourite month; I love the graveyard eeriness of the bare trees, the fungi, the ancient musty smell of rotting leaves and the bitter easterly winds. The black north-easter, the jovial wind of winter which fills the lakes with wildfowl, fills the marsh with snipe and hungers into madness, every plunging pike. Now is the time for fly fishermen mourning the end of the season to head for one of the Scottish lochs, such as Awe or Lomond in Argyll, Venachar or Acray in Stirlingshire, or any of the ones in the Great Glen and try for the next best thing — fishing for a pike with a fly.

Esox lucius are perfect killing machines, capable of fantastic bursts of speed; their long, sinuous bodies and head with eyes mounted on the top like a crocodile's are powered forward by a disproportionately large tail and dorsal fin set unusually far back. The monstrous mouth is filled with a staggering collection of teeth, containing incisors for gripping and immobilising prey, whilst the roof has hundreds of smaller ones pointing inwards towards a gullet capable of expanding like those of a snake. Periodically, this fails to stretch as much as a pike would hope and only the other day, one was found choked to death with the tail of a large carp protruding from its mouth. Pike are solitary predators relying on ambush as their main hunting method. Conveniently, nature has provided superb eyesight, sensors which can detect sound, vibrations and changes in physical pressure in the water around them indicating the presence of prey and given pike the extraordinary ability to change their skin colouring to suit both location and water at different times of year. In clear water the network of spots and blotches which cover the body turn a greeny-grey, to blend with reeds on the shoreline, or soil composition of the loch bottom. In peaty water, these become brown and yellow, or darker still if heavy rain brings down topsoil into their habitat and in floodwater, it turns to the colour of milky coffee.

Pike lurk immobile in the shadow of rushes at the water's edge, tree roots protruding from a bank, the timbers of a landing stage, or the stanchions of a bridge, waiting for prey to come within range and erupting out of hiding when one does. They are carnivorous opportunists, eating all other still water fauna dead or alive including smaller pike, from leeches, crustaceans and frogs to eels and the chicks of waterfowl. Pike can grow to 120 cm long and weigh 23 kg – with many historic records of much heavier weights; the 34-kg Loch Ken Monster caught in 1798 and measuring over 2 metres was dismissed as a fisherman's tale until one of 32 kg was found in 1934 at the mouth of the River Endrick, where it enters Loch Lomond. The head of the Endrick Pike is now an exhibit in the Kelvin Grove Museum in Glasgow.

More recently, one weighing 21 kg was caught in 2012 at Wykeham lakes near Scarborough and a year later, the head of a mammoth pike half a metre in length and 23 cm wide, containing 700 razor sharp teeth, some over 5 cm long, was discovered in reeds beside the River Cherwell in Oxfordshire. Fish of this size are a force to be reckoned with and there are innumerable incidents of pike attacking otters, dogs, cattle, horses and small children as they paddled in the shallows. In 1884, the renowned fisherman and author, H. Cholmondeley-Pennell, was severely bitten on the upper thigh by a forty-pounder, which subsequently chewed through the stick used to prise its jaws open and in 1999, a water skier was bitten on the foot by a monster which 'looked like a crocodile' at Llangorse Lake near Brecon.

Until the nineteenth century, when fly fishing became the vogue, pike were indisputably the most exciting fishing experience available. This was partly due to the possibilities of an enormous fish and the savage fight any pike puts up when hooked, but also the morbid fascination all anglers feel for a creature so prehistorically horrible, sharing the same waters as our gentle bream, dace, rudd and tench. Early authors of angling literature reserved special ingenuity for their advice on pike bait; Lauson in his *Secrets of Angling* (1620) suggests 'a younge whelpe or kitlin' (a young pup or kitten). Thirty years later, Izaak Walton in his definitive *Compleat Angler*, recommends a frog, preferably yellow and advises that it should be treated 'as though you loved him', whilst Cox in his *Gentleman's Recreation* of 1697, swore by a live gudgeon attached to a pinioned duck.

The modern fisherman uses dead bait, spins with a lure or trolls for pike, but this is dreary work to my mind, compared to one of angling's magic moments; the thrilling surge of power vibrating down the line when a pike takes a fly and tears off with it. All that is needed is a 2.74 metre, 10-weight rod, a reel with a good drag, enough spool capacity to carry at least 70 metres of 23 kg backing and line, a pike proof trace and some coloured flies. Pike do like a bit of colour. Oh – and something to jam in its mouth when you are removing the fly to stop it biting your fingers off.

Snipe

February, with its imperceptibly lengthening days, is a strange, unsettling time, more burdened with myths and folklore than any other. In farming circles, great significance is still attached to the weather on Candlemas day, 2 February. A fair start – if a badger can see its shadow – means there is more winter ahead than behind. The Saxons called February *Sprote-cal* (sprout kale), and below us in the valley there are the first signs of growth; a faint green haze of buds on alder, hazel and willow, with wild garlic and aconites pushing through along the riverbanks. In the hills, the first harbinger of spring is often the eerie melody of a cock snipe, announcing his arrival by performing a territorial display flight.

This hauntingly lovely sound puzzled naturalists for several centuries. One school of thought believed it was vocal. The other maintained it was produced by motion of the wings. The matter was not resolved until the 1950s, when Sir P. Manson-Bahr, R. A. Carr-Lewty and the then editor of *The Field*, Eric Parker, proved under laboratory conditions that 'drumming' is created by the two extended primary tail feathers vibrating in the airstream at a certain speed. The beautiful humming, which rises to a fluting crescendo, becomes audible when snipe, diving from a considerable height at an angle of 45 degrees, reach a speed between 40 kph and 65 kph. It only lasts for a few seconds and has extraordinary carrying and ventriloqual properties with the bird, disconcertingly, always some distance from where the noise is heard.

The soft bleating or whinnying which has given snipe a variety of colloquial names – *heather beater*, *horse gowk* and *mire kid* – seems to serve a number of purposes. It declares the cock bird's territorial boundaries and announces his presence to hens. Once one has been attracted, it forms part of the courtship display. During nesting, I believe the ventriloqual effect is used to distract predators. Cock birds arrive from low-lying marshes and wet fields at their moorland breeding grounds towards the end of February, the hens a week or two later. The main period of drumming activity is during mating and nesting, from mid-March until the beginning of July. Only the cocks drum and courtship is a wonderful exhibition of aerobatics as the birds chase each other, or preen and posture on the ground with tail feathers erect and spread like a blackcock's. Apart from their harsh scraping alarm call when disturbed, mating is the only other occasion when they become vocal, uttering soft eager cheeps.

The nest is a well-hidden, shallow scrape lined with dry grass in which the hen lays an average of four eggs, greenish with brown blotches. Incubation lasts about three weeks and while the hen sits the cock stays in the vicinity, displaying from time to time. The eggs hatch over a period of thirty-six hours and the first couple of chicks, which are active within a very short time, are hurried away by the cock to a nearby place of safety to protect the nest from predators. Both parents feed the young, which can fly at three weeks and feed themselves by the time they are a month or five weeks old. Snipe remain in the uplands until the first frosts drive them to lower ground. Their season is from 12 August until 31 January and the first snipe shot are often on grouse moors.

Common snipe (*Gallinago gallinago*), are mysterious little creatures of bogs and damp empty spaces of mists, water margins and rushes, mists and reeds, the rumble of surf and screech of sea birds. The term for a group in flight, a 'wisp', could not be more descriptive; seen briefly above a reed bed, they soon melt away into a grey winter sky. A number on the ground is called a 'walk'. They are artists at concealment and their striped plumage, the varying shades of yellow, black, light and dark brown and their long, pale-green legs are all the colours of marsh and estuary feeding grounds. They weigh no more than 0.17 of a kilogram and measure 25 cm to 30 cm, of which a quarter is beak. Snipe feed at dawn, dusk and moonlit nights on the small mud and slime dwellers that share their habitat, penetrating soft ground with their long beaks – a third of their body length. This long sensitive proboscis with its pliable tip enables them to locate their diet of worms, invertebrates, crane fly and other insect larvae. Although they usually probe for food, snipe will readily eat any surface or flying insect that comes within range, small molluscs and baby frogs. Their eyes are positioned high on their heads, enabling them to see upwards and ever alert to aerial predators whilst they feed. Response to danger is so instantaneous that people believed snipe used their beaks to propel themselves airborne.

Where suitable conditions prevail, snipe have an extensive distribution: north and central Eurasia from Iceland to the Bering Strait; most of Africa outside desert regions; and North and South America. Some of the finest shooting is in the rice fields of Carolina and the coastal swamps of Louisiana. Britain has a breeding population of around 70,000 augmented from August onwards by migrants moving south from Iceland, the Baltic and northern Europe, when the ground becomes too hard for their beak tips to penetrate. Large numbers often appear with the September and October moons along west coast estuaries and in Northern Ireland, before dispersing in search of suitable habitat.

The ideal wintering conditions are a combination of moist, frost-free soil and mud containing their food sources; some adjacent cover as protection from strong winds and predators; a bit of open water, which need be no more than the damp corner of a field where a blocked drain has flooded and freedom from disturbance. Enormous areas of snipe habitat have been lost, but the presence or absence of birds in any particular area is entirely dependent on their capricious response to the weather.

Apart from hard frost, which will drive snipe from their damp inland haunts back to the coast, their movement is governed by laws about which we know little or nothing. In some weather conditions they lie so tight you can practically tread on them; in others they consistently spring and jink away just out of range. In one season there will be any number at a marsh; the next year not one will visit. I was down at the Solway in late October on a day of blustery wind and heavy showers with wisps of snipe everywhere. The following day they had all disappeared.

Snipe shot in November, when they are feeding hard, are the most delicious eating and in my view, a 'finger of snipe' – traditionally, the number held between the fingers and thumb of one hand – are better even than woodcock. They are wonderful converters of food, coming to the table like little balls of fat and were considered beneficial for invalids, as the flesh was believed to be easily digested. This was possibly the reasoning behind Winston Churchill's reply; 'a finger of cold snipe and a pint of port', when asked for his favourite hangover cure. Like all game, its flavour seems to be enhanced in proportion to the effort expended. Shooting snipe is so utterly unpredictable and different from any other sport that it acquires a perverse charm all of its own, which easily becomes addictive.

Regardless of which continent they live in, most experienced sportsmen agree the only way to approach marshes or wet cultivated areas where you think they may be, is downwind. Snipe when flushed, rise against the wind and hang in the air for a fraction of a second, before flickering away. Like woodcock they zigzag from side to side, straightening out after a couple of dozen yards before gaining altitude. They do this to conquer the wind and control their power of flight, from then on, they accelerate hard. This, and the fact that you can never tell what distance they may rise from you — in some weather they lie so tight you can practically tread on them, in others, consistently get up just out of range — make hunting these elusive little people, both the most frustrating and occasionally, rewarding shooting.

Perhaps the greatest snipe-shot of all time was an American, J. J. Pringle, who abandoned a career in the US navy to devote twenty years to the pursuit of snipe through the bogs of Louisiana. Between 1867 and 1887 he claimed to have shot 78,602, before retiring to write what is still considered to be the definitive work on the subject, *Twenty Years' Snipe-Shooting*.

Britain's Alpine Heritage

In 2010, during the extended week from 5 February until the 14th, the town of St Moritz bounced spectacularly as the St Moritz Tobogganing Club celebrated the 125th anniversary of the world-famous Cresta Run and the long association between the sport and the town. A series of drinks parties, lunches, dinners and dances, guaranteed to weaken the strongest constitution were arranged over the ten days, plus a succession of extravaganzas. There was an aerial flypast and parachute jump, led by two indomitable Cresta members clutching banners carrying the Cresta logos; several motoring events, featuring classic motorbikes and cars, including a reverse Cresta hill climb, rallies, races and a parade of vehicles through the town. All proceedings were accompanied by bands, groups and soloists, ranging from a twelve-piece Bavarian brass ensemble to the inevitable enthusiast on the bagpipes. Cresta riders from all over the globe competed in eight races including the Grand National, the oldest and most famous Cresta race, which saw the 100th winner of this blue riband trophy. A ladies' race had been arranged for the first time since they were banned from riding in 1929; this one-off event is to be held every 125 years henceforth and George Bingham, the sculptor, was commissioned to create the Ecstasy Trophy, for the fastest lady rider. There were plenty of opportunities for letting off fireworks and the climax to the whole Olympian undertaking was the magnificent 125th Anniversary Ball.

The Cresta Run and indeed, all winter sports in the Alps, have their origins in a bet made in September 1864 by Johannes Badrutt, the proprietor of the Engadiner Kulm Hotel in St Moritz, to four English guests. For several decades, the pure air, sunshine and magnificent scenery of the Alps had been a popular summer destination for Britons seeking the thrill of mountaineering or, for those suffering from tuberculosis, an alternative to the steamy heat of the Mediterranean coast. The tiny hamlet of St Moritz had the added attraction of an average of 322 days sunshine a year, a dry invigorating climate and mineral springs renowned for their curative properties. At the end of the summer, as his guests were preparing to leave, Badrutt attempted to convince them that winter in the Alps was equally as attractive as the summer. If they returned in December, Badrutt promised them, they could stay as long as they liked at his expense and if he was wrong, he would reimburse their travel expenses. The four Englishmen had a glorious time lounging in the sunshine, skating on the lake, wallowing in the mineral baths and tobogganing on wooden 'schlittli', similar to an old-fashioned child's toboggan, returning home to spread the word that winter in the Alps was an experience not to be missed. Badrutt had won his bet, the Alpine winter season was born and with it, a unique and enduring Anglo-Swiss relationship.

Until skiing started in the 1890s, tobogganing was the principal entertainment and newly established winter resorts reverberated to the gleeful hoots and yells of Brits of both sexes, racing one another down the busy, winding streets of towns such as St Moritz, Davos, Aros and Chamonix. Before long, winter sports committees and tobogganing clubs were formed and regular competition races organised on the icy downhill roads. At one time, there were over forty 'village' pistes across the French, Swiss and Italian Alps, with the most challenging being the steep 3.2-km post road from Davos to Klosters. In 1883, John Addington Symonds, the British author and poet who was living in Davos for health reasons, founded the Davos Toboggan Club and organised the first International Toboggan Races. These took place over a week of balls and dinners, and it is indicative of the growing cosmopolitan attraction of winter resorts, that seven different nations were represented amongst the twenty-one competitors.

The success of this competition led to the Symonds International Challenge Cup and a little later, the Freeman's Trophy for women competitors, named after Miss Edith Freeman, a terrifying woman who rode her toboggan, *The Behemoth*, with total disregard for her own or anyone else's safety. Amongst the contingent from St Moritz, were five members of the British Winter Residents Outdoor Sports Committee: G. Robertson, Major W. H. Bulpett, J. Biddulph, C. Metcalfe and C. Digby Jones. Realising that Davos, the main winter resort to rival St Moritz, had stolen a march with their International race week, the committee decided to enlist the help of Peter Badrutt, Johannes' son, who was now running the much-enlarged Hotel Kulm, and build their own toboggan run. In autumn 1884, they staked out a three-quarter of a mile course following the natural contours of the valley from the Hotel Kulm, past the hamlet of Cresta to the outskirts of Celerina.

When the snows arrived in November, the committee set to work. Roger Gibbs, in *The Cresta Run 1885–1985*, describes them with their boots swathed in coarse bandages and arms linked, trudging time and again along the line that had been staked out until the snow was trampled down for the frost to harden. Creating earth banks to provide a framework on which to pile snow was a drawn-out affair and icing the run required endless buckets of water, but with labour and materials provided by Peter Badrutt, the run was completed in January 1885. The Davos Toboggan Club were invited to compete in the first Grand National on 16 February, which to the chagrin of the course builders, the visitors won, but the celebrations to baptise the Cresta Run in the Sunny Bar of Kulm eclipsed anything seen in Davos.

The original run was a rough and ready affair, down which both men and women careered sitting upright on their 'schlitti' and steering with their heels or short wooden picks, but improvements and innovations were quick in coming. Mr Cornish astonished everyone by lying head first on his toboggan to ride the 1887 Grand National, a position universally adopted, except by women, in 1888. That year a skeleton-framed toboggan was introduced with steel runners and a pad on which the rider lay, using metal rakes on his boots to brake and steer. This was further refined in 1902 with the sliding seat, which allowed riders to move their weight backwards and forwards as they negotiated the various banks. 'Traditional' toboggans are still in use today, although some of the more experienced riders use fixed seat 'flat-tops'.

By now the design of the Cresta was established as a serpentine ice tube of ten banked corners, approximately 1,212 metres in length with a drop of 157 metres and a gradient of 1 in 2.8 to 1 in 8.7, down which experienced riders hurtled at speeds approaching 112 kph, making them, at that time, the fastest thing on earth. The Run, which is open from just before Christmas until the end of February, is still hand built from scratch every year with the cost of building and operating the Cresta met by the St Moritz Tobogganing Club, its 1,300 members and active support from the St Moritz town council. It still follows the original route and is recognisably the same as the 1885 construction, although constant upgrading of bank design and icing quality now enables riders such as James Sunley, Christian Bertschinger, Johannes Badrutt, Count Luca Marenzi, the Gansser brothers or Lord Clifton Wrottesley to reach speeds of 128 kph.

The Cresta has survived into its first century and a quarter, partly because it is unique; there is only one Cresta Run and riding the Cresta is probably the last truly amateur sport. Partly because the Kuverein, the St Moritz town council, has always supported the Club through various financial vicissitudes. Partly because, although a private club, the SMTC have always been generous in providing time and expertise to encourage beginners, many of whom subsequently become devoted members. And partly, because the Cresta is an extreme sport that provides a thrill quite unlike any other.

In February 1973, I was in St Moritz for the last race of a long season's bobsleighing and in a rather blasé fashion, thought I would stay on for a week to try my hand on the Cresta. I remember my first ride far more vividly than any of the runs on various different bobsleigh tracks – even the one by moonlight on an old practice bobsleigh, with a pot valiant Frenchman at the helm. At 8 a.m., I was waiting at Junction, two-thirds of the way down the run where beginners start, clad in leather knee and elbow pads, gloves with protective metal plates, helmet, chin guard and spiked boots. Air Marshal Ramsay Rae, the SMTC secretary, announced over the tannoy that the run was free and with a series of ungainly kangaroo hops, I pushed the toboggan off.

Already things were going awry; I had landed too far forward and was badly out of alignment, banging from side to side as I picked up speed down Junction Straight, slithering round the first corner, Rise, like a drunken crab, hitting the wall an almighty thump as I came out. Battledore was negotiated rather niftily, but then the run dropped away and Shuttlecock, the long, raking left-hand bank, was upon me. Here I made the fatal error of peering about for the marker I was told to aim at to get the line for the bend, slewed dangerously to the left, dug my right foot in frantically and almost became unseated. Bounced round Stream Corner and wobbled down Bledisloe Straight, entering the sharp left hander at Bulpetts too high, earning myself another clout as I exited. Down the long straight to Scylla, round which I whizzed with surprising grace and on to Charybdis, with the ice rumbling inches below my face and an unbelievable impression of ever-increasing speed. Over Cresta Leap with my heart in my mouth to cannon painfully from side to side down the last stretch, arriving at Finish in time to hear the

Air Marshal announce laconically that J. Scott appeared to have arrived in one piece. As a bobsleigher, it was a deeply humbling experience, made all the more so when some of the Cresta riders later presented me with a cup inscribed 'Johnny Scott, Cresta King', but the abiding memory is of having achieved something remarkable.

Over the decades, the lure and excitement of the Cresta has attracted people of many different nationalities and backgrounds, a loyal following of millionaires and nobles, commoners and Royalty. Yet it has always been truly democratic. Membership of the SMTC is by no means exclusively aristocratic or even particularly wealthy. Indeed, two of the very best riders were St Moritz shopkeepers – the great Nino Bibbia, a local grocer, and Paul Felder, who owned a clock shop. From its very beginnings the St Moritz Tobogganing Club has been a partnership between the people of St Moritz and the British, and the town would not be the same without it.

History in a Wall Head

Most people dream of finding something of historic value, tucked away in an attic or lurking in the corner of a cellar. They live in hope that somewhere in their home, overlooked or discarded by previous generations, they will make a fascinating discovery. For some, this eternal optimism turns into an obsession. My brother-in-law, for example, persuaded himself that a well, situated in the oldest part of his house contained coins tossed there over the centuries for good luck. Gold sovereigns could possibly be among them, and the only way to find out was to go down the well with an aqualung and oxygen bottle lowered behind on the end of a rope. I felt deeply for him when the mud at the bottom contained nothing but rat skeletons, broken champagne glasses and the handle off a chamber pot, because I too, was once bitten by the bug.

In my case, it happened when I was very small. In those days, it was common practice for certain members of the local hunt to tow wagons to the point-to-point field, a couple of days before the event. These were usually old, high-sided horse-drawn carts with the shafts replaced by a metal tow bar. Lined up in a row, they were wonderful places for entertaining friends. Their elevation gave a good view of the proceedings and the high sides stopped people falling off. Ours was a big hay wain and I was allowed to ride in the back as it was towed to the field. This might have been excitement enough for a child, but it was nothing compared to the thrill of finding a small, heavy object wrapped in mouldy oilskin, lying on the wooden floor. Unwrapped, this proved to be a single-barrelled, percussion cap pocket pistol with the maker's name, Smith of London, just visible through the rust.

The pistol must have lain on a beam or wall head of the cart shed for over a hundred years and been knocked off when the heavy hay wain was moved. Endless speculation between my father and the farm men about who it had belonged to or why it had been hidden there, convinced me that if you wanted to find treasure, old farm buildings were the place to look. For a long time afterwards, I searched for whatever happened to be the childhood fantasy of the moment – usually firearms related – crawling along beams and clambering up to peer into the space where the wall heads joined the roof. As I grew older and my interest in the land developed, searching for the improbable was replaced by something infinitely more rewarding.

British agriculture changed out of all recognition during the 1970s and 1980s. Traditional mixed farming, with its rotation of grain, roots and grass became a thing of the past. The patchwork landscape was converted to large field monoculture as hundreds of miles of hedges and acres of small woodland was bulldozed out. Fifty per cent of the workforce left the land as technology replaced manpower and when tractors got bigger, conventional stone and brick-built farm buildings were considered obsolete. Thousands were demolished or converted to homes and office units, to be replaced by portal frame structures of breeze block and pressed steel.

What I find so fascinating are the little insights to our farming heritage that are waiting to be found in those that survived. Our dressed sandstone and slate-roofed steading, was built in 1823 and is still standing. The feed store, stables, cattle byres, cart shed, granary, hayloft and various outbuildings were all in continual use until the 1980s. Generations of farm workers used the wall heads, where beams or roof joists formed convenient alcoves, to store anything small enough to fit them. Overlooked when the buildings were cleared, it is here that I find my gems.

Investigating wall heads is a splendidly filthy occupation best done on a cold winter afternoon. Each part of the steading, depending on what it was used for, will reveal something entirely different. Over the years, I have developed a nose for where to expect objects of interest. The feed store has never provided anything, because it was a dirty dusty place to work in and people spent as little time in there as possible. In the cattle byre, on the other hand, I have found a variety of narrow four-sided glass vaccine bottles, part of a set of leather casting hobbles for immobilising cattle. A wooden gag with canvas straps and a balling gun – a copper tube with a rod inside it, for pushing a mineral bolus down a calf's oesophagus. The curved blade of a hoof-trimming knife, cemented by rust in a tangle of chain and a set of bull tongs with rigid handles, for restraining a bull by the nose. The best find, buried under years of dust, was a cow's horn with the end cut off, for administering a liquid vermifuge to cattle with worms.

Lying across the beams in the cart shed are the wooden swingle-trees and short back straps that would have been part of the harness for a reaper. Beside them are the serrated cutting blades, packed between planks of wood and the sides off a single-axle farm cart. Length wise on the wall head, I found a pair of elegant, mauve and black painted shafts from a fancy pony trap and the broad, heart-shaped head of a 'fauchter' spade, used in the laborious process of digging peat for winter fuel. Next door in the stables and adjoining tack room, there is any amount of miscellaneous carthorse harness – saddles, collars, thick leather trace lugs, belly bands and bearer straps. The wall head was a convenient place for standing breakables out of harm's way and there are still stoneware jars of Clark's cracked heel liniment. Bottles of Elliman's Embrocation – guaranteed to cure everything from a sprain to a cough and a tin of Gombault's Caustic Balsam, rotten with rust.

Up above is the hayloft, where the farm men ate their 'piece' on wet winter days – I can see them in my mind's eye their heavy hobnailed boots with the turned-up toes, leather leggings and thick, Derby tweed breeches. The wall head here has given up all manner of curiosities; empty beer bottles, one with a faded label that read: 'Younger's Sweetheart Stout'. A soft-drink bottle with a glass ball in the neck to stop the gas escaping and another, with the glass designed in a series of art deco swirls. A variety of Wedgwood-patterned plates in pink and blue. A stoneware jar from Harvey's Creamery, Ayr and a short-stemmed briar pipe with a perforated metal lid to stop hot ash falling out.

Perhaps the most intriguing discovery was in the bare little room at the end of the loft, where a single man would once have lived. It has always seemed such a lonely place and the word 'welcome' painted on the door, rather tragic. Poking about on the wall head above the truckle bed I found a red button, shaped like a flower with a piece of lilac thread attached. Now whatever can that have been doing there?

Starlings

Before the odious Hunting Act 2004 banned hunting with dogs, the famous Cholmondeley Coursing Club had its headquarters at the Cholmondeley Arms near Malpas in Cheshire, where the numerous trophies won at coursing meetings over the years, had accumulated above the bar. Members based themselves there for the Waterloo Cup, the blue riband of the coursing season, travelling back and forth from Altcar over the three days of this historic sporting event. Our route took us across the Runcorn Bridge and a sight which became synonymous with the cup, were the vast flocks of European starlings, as many as a hundred thousand, wheeling in the late February dusk above the Mersey, before streaming down to roost on the bridge girders with a deafening cacophony of exultant squeaks and whistles.

These 'murmurations' or 'starling skies', made up of thousands of birds wheeling and swarming like bees above the landscape, swelling with ribbons of late arrivals rushing from their feeding grounds to join the main body, are one of nature's great phenomena. A display of high-speed synchronised aerobatics on a massive scale, which is quite breathtaking to watch. Until the 1970s, when the population reached its peak, vast flocks of domestic starlings and migrants from Scandinavia and the Baltic were a familiar sight on winter afternoons, smothering telephone and electricity wires or swirling like some amorphous organism above a favourite roost. These, presumably because they provided perching space to accommodate the ever-increasing numbers, tended to be urban with massive numbers of starlings pouring in at dusk to roost in all major towns and cities.

When I think of Birmingham, Manchester, Leeds and particularly, the West End of London in those days, I remember the never-ending cacophony of frenetic birdsong, buildings and statues streaked with dung and foul slippery pavements.

By now, starlings were considered to be a major pest species. Their droppings are acidic enough to erode stonework and health authorities became alerted to the infection risk from histoplasmosis, a lung disease caused by a fungus that forms on bird dung. There was further concern over the enormous number of mites generated in starling nests, which are notoriously filthy. Borough councils up and down the country embarked on starling eradication programmes which also led to the control of the other avian pest species, sparrows and feral pigeons. Nesting sites were blocked off, and initially, buildings covered in unsightly wire or plastic mesh to prevent starlings from roosting. As businesses specialising in bird control improved techniques, these were replaced with monofilament wires stretched across ledges or boards set at angle of 45 degrees, mats covered in plastic spikes and repellant gels. One of the most successful, used on runways, was the development of the distress call scaring device. An audio tape of an anguished starling is broadcast either through hand-held megaphones, or fixed amplifiers. Believing they are under attack from predators, roosting birds will vacate the area.

At the same time, their principal food source began to diminish dramatically. Starlings are omnivorous and will eat practically anything. In the autumn, large flocks are deadly to fruit growers and can cause significant damage to arable crops by digging up germinating seed, but 50 per cent of a starling's diet is made up of any insect and in particular, earthworms, wireworms and leatherjackets. The nutritional value from these is essential for fledgling survival and a starling's beak is especially equipped with highly developed protractor muscles, which enable it to pry open the ground in search of their main prey items. During the 1970s and 1980s, agricultural pesticides and the reclamation of old permanent pasture – over a million hectares disappeared between 1970 and 1995 – caused the domestic starling population to drop by 56 per cent over twenty-five years with an equivalent effect on winter migrants.

Despite these deprivations, European Starlings remained one of the world's most populous birds. They are native to Europe, North Africa and Asia with the northern European population migrating south to winter in the Mediterranean, Africa, Egypt, Iran and India. During the late nineteenth century, they were introduced to North America, New Zealand and Australia either by sentimental British colonists or to control insect pests. The ancestors of the North American population, now estimated to number 200 million – a third of the world total, were introduced by Eugene Schieffelin, the eccentric son of a prosperous New York lawyer. Eugene was an ardent fan of Shakespeare and thought it would be nice to import all the fifty-six bird species mentioned in Shakespeare's works and release them in Central Park. The sentence in *Henry IV* where Hotspur proposes tormenting the king by '… I'll have a starling shall be taught to speak nothing but "Mortimer" and give it to him to keep his anger still in motion', caught Eugene's eye and in 1890, sixty pairs of starlings duly arrived from Britain. Fifteen survived to create what has become an ecological nightmare from arctic Canada to subtropical Mexico and directly led to the US laws banning the imports of alien bird species.

Furthermore, starlings have a thick-skinned personality ideally suited to an invasive pioneer. They are the delinquents of the bird world – garrulous, aggressive, self-confident and ruthlessly opportunist.

Individually, they are great comics, mimicking the sounds of other birds and some animals for their own amusement. Caught young enough they are easily tamed and were often kept as pets for their ability to learn tunes and simple phrases. En masse, they rely on numbers, noise and the sheer energy of their rapid flight, to bully other birds off feeding grounds and away from nesting habitats. Even their mating is loutish. Cock starlings make a cursory nest of twigs, straw and leaves in gutters, drainpipes, wall cavities or the nests of others — woodpeckers, swallows or martins, sometimes evicting the eggs and fledglings of the original occupants. His normal incessant chattering call turns to a mellifluous whistling which he performs with drooping wings and puffed-up throat feathers. When this pathetic display attracts a female, his song becomes strident and demanding whilst he hunches his back, beating his wings. If his message has not been clearly understood, he clambers into the nest itself, where both song and wing beats become frenzied. The hen is left with the task of lining the nest and making it habitable. Although some cock birds may help incubate the eggs for a short period during the day, most tend to be polygamous, abandoning the hen soon after mating. Hens lay five to seven light-blue eggs which take twelve to fifteen days to hatch and a number of starlings nesting in the same area will synchronise hatching. Their population spread is accelerated by an extended breeding season which lasts from April to July and the hen bird's ability to rear three clutches of eggs, if the weather remains favourable.

By the 1990s there was such national concern over the decline in our domestic population that starlings qualified for inclusion on the Red List of Species of Conservation Concern. Since then, due to a complete turnaround in government agricultural policy, Set Aside schemes, the growth of organic farming initiatives, Countryside Premium and Rural Stewardship Schemes have combined to replace much of starling's natural food source and their numbers are on the increase. The UK breeding population now stands at around 9 million and at this rate, 'starling skies' will once again be a familiar part of our winter countryside.

Stoats

Stark December weather this week, with a gliff of snow on the tops and gravelly ice glinting on dead bracken fronds in the pallid winter sunlight. The hills are as grey as the sky and there is a leaden, brooding feel to the atmosphere, raising hopes among the young for a white Christmas. They might be lucky – the hill sheep, always a dependable indicator are hanging about in groups as if they know something. Coming down off the last bit of high ground, I heard the thin hopeless screaming of a rabbit. Below me, where a hollow in the hills broadens out to a track leading back to the steading, are the tumbled remains of a circular stell and an old railway goods wagon, used as an emergency hay shed. Hay seeds trodden in over many decades, have created an improved sward enjoyed by the rabbit population from a warren on a bank nearby and at the bottom of this a stoat was entwined round one, head wrenching savagely as he sought to bury his teeth into the base of the skull. Satisfied the rabbit was dead, the stoat peered rapidly round in every direction, muzzle lifted and ears pricked before darting off into a clump of rushes.

Within moments he had returned to the carcase and began dragging it backwards in a series of frantic tugs towards the hay shed, his long body coiling and uncoiling in spasms of effort. Periodically he stopped briefly, not to rest but to sit upright on his haunches, nostrils twitching for scent before continuing his journey. Stoats are normally crepuscular, relying on their acute hearing and sense of smell to alert them to danger and it was indicative of hard weather that he was hunting in broad daylight. At the goods wagon, he slid through a gap where the corner of a plank had broken off and in a series of heaving jerks, the rabbit disappeared after him. An incredible feat of strength when you consider that the rabbit weighed six times as much as the stoat and his labours were by no means over. The hay in the goods wagon is only fed when sheep are driven down by a heavy fall of snow and as I knew from experience, the top layer of bales would invariably be covered in a disgusting mess of rabbit carcases and their digested remains.

Part of the Mustelidae family, which includes otters, badgers, pine martens, polecats and weasels, stoats are widespread throughout the British Isles, northern Europe, Scandinavia, North America and temperate regions of Asia. They were introduced to New Zealand in 1884 to control the rabbit population and were subsequently the primary cause of the extinction of several rare bird species and decline of others. Stoats are the most perfect, utterly fearless, opportunist killing machines. The embodiment of sinuous agility and determination who, in the words of *The Irish Sportsman*, 'runs like a fox, dives like an otter, climbs like a cat, doubles like a hare and will stand before hounds for hours and beat them'. Their diet is sufficiently varied for them to have survived the loss of their principal food source through myxomatosis. They will hunt anything from shrews to hares, including reptiles, amphibians, fish and even, *in extremis*, berries, earthworms and insects. They are deadly to ground-nesting birds or in a hen run, can shin up and down trees headfirst like a squirrel and cover the ground in an effortless lolloping gallop at a speed of 20 mph. Such is their reputation in the animal world for relentlessly holding to scent, that rabbits and hares become paralysed with fear when they know they are being hunted by one, pathetically waiting for death. The rabbit just killed had been within yards of the safety of a burrow to which it would have bolted at the first sign of any other predator.

The goods wagon and stell provide ideal stoat habitat. There is plenty of prey in the surrounding neighbourhood and the jumble of stones, a convenient nesting site to give birth in. Bitch stoats mate in July or August, often with multiple partners and have a gestation period of eleven months, the longest delayed implantation of any mammal. She rears up to twelve kittens single handed which are born blind and covered in white fur with a thicker patch at the nape, used by the mother if she needs to carry them. The kittens open their eyes after a month, are weaned at five weeks and the distinctive black tip to the tail, which identifies them from weasels appears at six. It is not unusual for an adult dog stoat to force his way into the family group and inseminate a juvenile bitch at this age. After weaning, the family stays together for some time and I have occasionally watched a pack of young stoats in the vicinity of the goods wagon, being taught to hunt small mammals by their mother and once saw a young rabbit with a bitch stoat and six young clinging to it. Like all mustelids, stoats are inquisitive, playful and natural comics. A family at play, chasing each other at terrific speed, deliberately running to the top of a bank and tumbling down or collapsed in an exhausted heap is one of the great sights of nature.

In late the autumn young disperse to find their own territories, establishing several dens in abandoned burrows, rock crevices or among a pile of stones. They now become solitary, meeting only when the bitches are in oestrus. Both sexes fiercely defend their hunting grounds that range from 2,000 to 6,000 square metres, depending on the type of habitat and food source. These are marked with pungent scent from the glands on either side of their anus. Like polecats, pine martens, ferrets and mink, stoats will use their scent glands as a powerful deterrent if attacked by a large animal that knows no better. Many a foolish young dog has undergone the indignity of repeated bathing until the appalling stink wears off.

Except for the glossy black tip to their tail, stoats turn white in winter. The rich copper colour of their summer coat disappears over a couple of days as the creamy belly fur spreads upwards over the back. My little hunter was in the half pelage, white with a patchy brown band from his head to his haunches. Unlike blue hares, which turn white seasonally regardless of temperature, a stoat's pelage changes colour climactically. Some stoats seem more inclined to adopt winter clothing than others, with white stoats being seen during a temperate winter in the south. In parts of the Highlands, they remain white all year round. During the Little Ice Age, ermine as the white fur is called, became an extremely valuable commodity and wearing it, restricted under medieval sumptuary laws to the highest ranks of nobility. Ermine imported from Russia and later, Canada, then became the official mantling of coronation and parliamentary robes worn by peers. It took up to thirty skins to create the heraldic white field flecked with black tales. To ensure that the peers were properly turned out, 50,000 ermine skins were imported from Canada for the coronation of King George VI, in 1937. Nowadays, they are made from artificial fur and these lovely pelts can be bought for as little as $14.

Curiously for an animal so merciless, a white stoat was once considered in Europe, to be a symbol of purity and chastity, which would rather die than get its coat dirty. Leonardo da Vinci's famous painting, *Lady with an Ermine* (1483) which hangs in the Czartoryski Museum, Kraków, is an ironic play on this belief. His subject Cecilia Gallerani, depicted cradling an ermine, was the seventeen-year-old mistress of the Duke of Milan, whose heraldic design was an ermine.

The Golden Queen of the Woods

Out on the edge of the moor, the ruins of a shepherd's bothy nestles in a sheltered south facing valley where a small burn meets a bigger one. It is protected from the north by a windbreak of old Scots pines and contorted larches and on the other three sides by the steeply rising slopes of the surrounding hills. Nearby, in a basin of flat ground, the earth walls of the shepherd's garden can be seen, where rhubarb planted perhaps 200 years ago, still grows every year. Climbing a sheep path above the ruin, a woodcock, the first I have seen this year, suddenly materialised out of a patch of dead bracken where it had been dozing in the sunshine and, after jinxing from side to side, silently curved away to fall among the trees.

I had half expected to see one of these magical birds, whom the French so eloquently call *La Mordorée* – the Golden Queen of the Woods. Wildfowlers over on the Northumbrian coast had reported the first wave of migrants from northern Europe, driven south when the ground became too hard for the pliable tip of their beak to penetrate, lolloping like great moths under the Hunter's Moon around Halloween. Exhausted birds have also been seen with their feathers puffed up, recuperating in the bent grass among the sand dunes. It is extraordinary that a bird completely unsuited for prolonged flight, whose digestion is so rapid and intake of food, normally so frequent, should undertake a long sea crossing. Success depends on the north easterly wind and disaster lies in being caught in an offshore blow that prevents them from landing. The extremes in condition between woodcock that have had an easy or difficult crossing led people to believe that there were two distinct species. With a plentiful supply of food, their voracious appetite leads to a rapid recovery and with open weather in early November, woodcock often stick to high ground as they move across country before finding mixed woodland with decent undercover and rides or open glades to winter in.

For sheer fascination, woodcock stand shoulder to shoulder with brown hares and on the rare occasions when they can be watched undisturbed, always provide a thrilling and memorable experience. The silent, ghostly movement along a woodland ride as one makes its lazy journey to a favourite feeding ground at dawn or dusk. The way the dumpy little creature stands four-square on the edge of a damp clearing, probing the ground with long delicate beak, occasionally performing a rapid stamping dance to bring worms closer to the surface. The spring migration starts in February or the beginning of March, depending on the weather with woodcock filling the woods along our east coast, waiting for the right wind for the return journey. This is when our otherwise mute resident cock birds become vocal as they perform their territorial mating flight, known as 'roding'. Flying at tree canopy height they quarter their territory, periodically uttering a series of far carrying frog-like croaks ending in a high-pitched squeak, to which the silent female responds by exposing the white underside of her wing. Unlike most birds, woodcock are philanderers, staying with the hen for only a few days before searching for another. Roding can continue until August as immature and less experienced males search for available females. Nests are rarely seen, with the hen relying on her superb camouflage to integrate into a background of understory for protection. She raises her brood of four chicks single handed and every countryman longs to see one of them transporting a chick in flight, clutched between her thighs.

Years ago, when I was a mud student on a farm in East Lothian, living alone off a diet of fry-ups, I found a woodcock lying among the leaves beside a path leading to my cottage. It appeared to have been spooked into the telephone wires and although still warm, seemed completely lifeless, with head and neck dangling loosely as I held it in my hands. Woodcock are undoubtedly the finest eating of all game birds and I was torn between gastronomic anticipation and sheer wonder at the perfection of its plumage. All the shades of autumn – yellow, black, chestnut, russet, citron, gold and copper, perfect camouflage for a bird whose preferred habitat is brambly woodland with plenty of ash and sycamore. Whilst I was lost in admiration, the neck eerily rose up and the bird wriggled into a sitting position, head groggily wobbling from side to side. The eyelids flickered open and the disproportionately large, high-set black eyes gazed blearily round. Before I could react, supper had taken to the air.

Woodcock have always been appreciated for their exquisite flavour – the great culinary encyclopedia, *Larousse Gastronomique* lists more than thirty different recipes, but until the eighteenth century they were considered incredibly stupid birds for the ease with which they could be netted. They generally follow the same route to their feeding grounds and with a little research, a fine net suspended across a ride or the entrance to a woodland clearing would guarantee a catch. This and the effortlessness with which exhausted migrating birds were caught on landing, fuelled the view commonly held for many centuries, that woodcock had no brains and the name was synonymous with idiocy.

Improvements in shotgun design from the 1800s, which made wing shooting feasible and the bird's elusiveness, scarcity and aerial dexterity in woodland when flushed, elevated woodcock to prominence as one of the most sporting and sought-after game birds across Europe, Canada and America, where a subspecies, *Scolapax minor* – otherwise known as the mud sucker or timber doodle – is found from Newfoundland to the Gulf of Mexico. There is nothing, in any language, quite like the shout of 'mark cock' to set a gun on his toes. In France, shooting woodcock over pointers is almost a religion and the degree of fascination the French have for this strange little bird is reflected in the volume of literature written about it. Colin McKelvie, who has made a study of woodcock, mentions ninety-two popular titles and over 2,000 scientific publications, to say nothing of endless articles in the sporting press.

Most people are so chuffed to shoot a woodcock that they pluck the pin feather from the elbow of either wing and treasure them as mementos. Some display the feathers in their hatband, others like me, Sellotape them into game books and on the Continent, they are often buried with a favourite pointer. These small, sharp pointed feathers were once in demand by artists for delicate paintwork. Medieval illustrators used them to illuminate the panels of ecclesiastic manuscripts and Bibles, whilst the Victorians used them for painting exquisite miniatures. One of our leading wildlife artists, Colin Woolf, began to use the same technique twelve years ago, to paint the most beautiful pictures of woodcock. One of these, auctioned for charity in 2000, was painted with a 150-year-old feather found in a paintbox belonging to the miniaturist, Lady Letitia Kerr.

According to Colin, in the Far East, where woodcock are no less revered, pin feathers mounted in a delicate gold or silver tube are used as a sex toy. Now there's an idea for a Christmas present.

Coppicing

Broadleaved woods are quiet places in winter; havens of peace and tranquillity. There is an old oak, ash and beech wood, with hazel and willow where it drops down to a stream near here where I go and sit sometimes and listen to the wind sighing through the treetops or rustling the leaves on the ground. Once upon a time it was a working copse and these trees would have been coppiced – copse comes from the French *'couper',* to cut – and if I let my mind drift, I can see the wood when it throbbed with the sounds of woodsmen working. The rasp of a two-handed saw as sawyers drew it back and forth, laboriously sawing a big standard tree into planks in a saw pit especially dug for the purpose. There is the 'tock' of axes, the clink of hammers striking wedges and the sweet smell of woodsmoke; the snorts of a carthorse leaning into its traces and the creak of wagon wheels. Beside the sunken track worn by centuries of timber wagons, are piles of different-sized firing wood for brick kilns or charcoal; faggots, billets, bavins and stackwood. There are cutting blocks, cordwood, planks, long ash poles, fencing posts, willow whips, twigs and bark for tanning.

Coppicing is the oldest form of silviculture, practised from the time early man first discovered that having laboriously cut down a tree, the stump produced a self-renewing supply of timber and it is still a method of conservation management in broadleaved woodland. The earliest archaeological records of coppicing in Britain were discovered in 1970, when peat diggers unearthed part of a wooden walkway, the timbers of which have been carbon dated to 3900 BC. The Sweet Track, named after its discoverer Ray Sweet, extended across the waterlogged marshes between an island at Westhay and a ridge of high ground at

Shapwick, a distance close to 2,000 metres. The track is an elaborate structure, engineered from coppiced poles of ash, oak and lime driven into the marsh to support a walkway that mainly consisted of split oak planks laid end-to-end. The Sweet Track and archaeological remains of Neolithic hut construction clearly indicates there was an existing historical culture of coppicing ash, oak, hornbeam, lime to provide straight poles of about 5 metres for structural supports, with willow and hazel rods for wattle and daub walling.

Coppicing was brilliant in its simplicity. It allowed a woodland crop to be harvested with simple hand tools and in manageable sizes and weights, whilst always self-regenerating to provide a continual source of material; there was no necessity to clear-fell woodland except to create farmland. Archaeological evidence demonstrates that coppice products were used for innumerable rural needs throughout the Bronze Age, Iron Age, Roman and Saxon periods. During the 400 years of their occupation, the Romans required an immense quantity of wood and much of Britain was still covered in wildwood forests. In fact, had it not been for the availability of this essential raw material, it is doubtful whether the gold, silver, tin, copper, lead, iron and other commodities which Britain had to offer, would have been sufficient inducement to invade. They introduced a system of 'coppice and standard', where individual trees were allowed to mature among coppice, as timber for bridges, the great urban buildings, villas of the rich and ships of the navy. Roman mining operations were on an industrial scale not seen again until the sixteenth century; the historian Dr Oliver Rackham, has estimated that over 23,000 acres of coppiced wood was needed for charcoal to fuel the military ironworks in the Weald alone.

Coppicing remained the most widespread method of woodland management until the mid-eighteenth century and preserved our ancient woodlands intact. Except for ornamental planting on the great estates to enhance the landscape or provide cover for game, virtually no trees were planted. There was no need to; the coppice and standard system continued to work perfectly well and in the north of Scotland, the Scots pine forests stretched into infinity. The idea of planting specifically for timber only occurred in the late seventeenth century and was largely influenced by the publication, in 1664, of *Sylva, A Discourse of Forest Trees and the Propagation of Timber in his Majesties Dominions* by John Evelyn, the diarist, horticulturist and founder member of the Royal Society. Evelyn had been horrified by the wanton destruction of woodland during the Civil War and the mismanagement during the years of Cromwell's Protectorate, when many Royalist landowners, such as the Reresbys of Thrybergh, in the West Riding of Yorkshire, were forced to destroy extensive stands of hardwood to pay fines imposed by Cromwell. Evelyn believed, erroneously, that woodland in Britain was in terminal decline due in part to the deprivations of the previous twenty years, but also from the demands of various industries, such as iron, house and ship building.

A Discourse of Forest Trees was addressed primarily to the Lords of the Admiralty as a warning of impending timber shortages for the navy and advocated an immediate policy of woodland planting. It is surprising, considering his fortune was based on charcoal and family gunpowder mills at Godstone, Wotton and Tolworth, that Evelyn failed to recognise the historical evidence that industry had been responsible for sustaining our woodlands, rather than destroying them. Nor did he appreciate that felling broadleaf trees does not kill them and provided they are protected from browsing animals, will regenerate in a matter of years. Throughout history, nearly all woodland clearance had been for agriculture and up until the Industrial Revolution, industries relied on charcoal from coppice woods for fuel.

The great woodlands of the Weald had already been supplying fuel for the local iron industry for a thousand years before its heyday in the sixteenth century, when over fifty blast furnaces and sixty forges were churning out cannons for the Tudor wars and could never have survived unless they had been managed as coppice. The same was true of the other mining areas, such as the Merthyr and Ebbw valleys in Wales, or the Forest of Dean. It was in the agricultural areas such as East Anglia, the Midlands, the flatlands of North Humberside, the Vale of York, or the coastal lowlands of Scotland where woodlands were almost completely destroyed. To quote Dr Rackham, 'the survival of almost any large tract of woodland suggests that there has been an industry to protect it against the claims of farmers.' *A Discourse of Forest Trees* influenced many landowners, including King Charles II, and is now regarded by historians as being responsible for much of the disinformation about trees that is still current today.

Contrary to the accepted policy of coppice and standard mixed woodland, Evelyn advised landlords to make extensive new plantations of only one or two species together. Beech was planted extensively in the eighteenth and nineteenth centuries; the beech plantations in the Chilterns are an example of traditional coppiced woodland being sacrificed to the needs of the furniture industry. In other areas, landlords increased their oak woods or followed the fashion for sycamore, poplar, wych elm, hornbeam or conifers. Evelyn had a passion for 'exotics', as conifers were referred to and considerable quantities of Norway spruce, silver fir and European larch were planted in the eighteenth century. Between 1738 and 1830, successive Dukes of Atholl – the 'Planting Dukes' – planted 27 million conifers, mainly European larch from the Austrian Tyrol, on the bare hills of their estate at Blair Atholl. The fourth duke was so obsessed with plastering the landscape with trees that he established conifers on the inaccessible and almost perpendicular slopes of Chreag Bhearnach, a jagged mountain overlooking Dunkeld, by firing canvas bags of seed at them through a cannon.

If landlords following Evelyn's advice fuelled the decline of coppicing, so too did the Industrial Revolution as coal or coke replaced charcoal and many historic natural woodlands disappeared. Far more damaging were the forestry policies before and after the Second World War which implemented conifer planting in northern England, Wales and Scotland on a massive scale. Vast areas of beautiful, wild open spaces became filled with dark, regimented, forbidding conifer plantations, displacing isolated farming communities and engulfing many semi-natural woodlands. I remember being taken as a child by my father on one of his monthly visits to the family farms in Northumberland, to watch the planting of a big area of hill above Stannersburn and thinking how strange it looked, as a caterpillar tractor dragging a huge Cuthbertson forestry plough gouged black lines through the green fields in front of an abandoned upland farmhouse. Today the Forestry Commission manages 7,720 sq km of land in Great Britain; 60 per cent is in the hills of Scotland, particularly the Highlands, western Borders and Galloway; 26 per cent in England including Kielder Forest, which covers 650 sq km, with the remainder in Wales. It has been estimated that in the thirty years between 1945 and 1975, nearly half the remaining ancient woodlands of England, Wales and Scotland were seriously damaged or destroyed; more than in the whole of the previous 1,000 years.

Active commercial coppicing survived in a small way through the twentieth century, mainly in the sweet chestnut – introduced by the Romans – coppices of Kent and East Anglia, with the principal outlet being the fencing industry. In the last decade the wheel of history has turned slightly and there has been a revival of coppicing, especially of hazel in Hampshire and other southern counties, oak in the north west and beech for bodging — making furniture legs — in the Midlands. This is due to conservationists appreciating the importance of coppicing in maintaining traditional ancient woodlands, and the growth of interest in traditional crafts. A new generation of coppice workers and woodsman have developed markets for chestnut paling, wattles, trug baskets, faggots for riverbank stabilisation, barbecue charcoal, greenwood furniture and garden ornaments. It is encouraging to reflect that the demand for good quality coppice now exceeds the supply.

Our Fathers of Old

February is a wretched month in the Borders of Scotland, best described by that marvellously onomatopoeic Scots word, '*dreich*'. A dreich situation is one that has become dreary, unpleasant and drawn out. A dreich day is a miserable one; foggy, chilly and damp. By now the winter has become detestable, hanging on for far too long. Everything around is dead; the hills of Liddesdale are grey and lifeless; the trees gaunt and leafless; sunlight is thin and watery; the ground cold and wet. Towards the end of the month, a change occurs, and under the alder trees that grow along the banks of the Hermitage Water, a beautiful green living carpet of spear-shaped leaves emerges. These are wild garlic, *Allium ursinum,* and their appearance is greeted by us with the same enthusiasm and excitement as that felt by generations of our predecessors, long before the Novantae Celts built a fortified village overlooking the valley.

Ramsoms, buckrams, wood or wild garlic (as they are also known) are the first wild growth of spring and were once a longed-for source of fresh food, minerals, trace elements and vitamins after months of eating salt meat and dried vegetables. The plant produces a profusion of brilliant white, star-shaped flowers in late April or early May and could be mistaken for lily of the valley but for the overpowering, pungent, garlicky scent which on a warm day can be unmistakably recognised from yards away. Cattle find them irresistible and the milk from dairy cows grazed on old pasture where wild garlic grows, immediately becomes tainted by it. For centuries, the leaves were made into soups and purées or used as flavouring for stews and sauces. John Gerard, in his *Great Herball* of 1597, mentions the use of buckram leaves in a sauce for fish and observed, 'they may very well be eaten in April and May with butter, by such as are of strong constitution, and labouring men'. We sometimes eat them like this to accompany steak, softened in a pan with butter and the meat juices.

There are literally hundreds of different wild plant species in Britain and virtually all of them, even the poisonous varieties, were once extensively used by both rural and urban populations. An enormous number were edible, either fresh or preserved, and were depended on by isolated rustic communities in times of drought or famine. Others were harvested for their curative properties and for many centuries constituted the only known source of available medicine. Many plants were dual purpose, being edible and containing remedial properties at the same time. Still more had an infinity of practical uses; reeds, rushes and marram grass for baskets, bee skips and thatching; bracken for bedding stock or packing breakables for transport. A cottager in the eighteenth century, whether living on coast or estuary, mountain, moorland, flat or fen, would have an encyclopedic understanding of the wild plants and herbs that grew in his district.

Even after the enclosures and agricultural improvements of the 1800s, which brought prosperity and an abundance of available produce to most areas, wild foods were still assiduously harvested. This was especially so in early spring, when winter stores were running short and cultivated crops were yet to grow, or during periods of localised failed harvests. It was on occasions like this that wild foods came into their own; hardy, resilient and able to thrive on the poorest soils, our indigenous plant life has always been there in times of need. Dependence on nature's bounty of plants and herbs gradually diminished as industrialisation drew people away from the land and when medical science made great strides forward in the development of mineral-drug-based pharmaceuticals. Although herbal remedies were still used in rural areas to treat all but the most serious illnesses, their efficacy was sneered at elsewhere as ignorant folklore. Kipling's poem, 'Our Fathers of Old', reflects the view of the intelligentsia in 1910, part of which I enclose:

> *Excellent herbs had our fathers of old –*
> *Excellent herbs to ease their pain –*
> *Alexanders and Marigold,*
> *Eyebright, Orris, and Elecampane –*
> *Basil, Rocket, Valerian, Rue,*
> *(Almost singing themselves they run)*
> *Vervain, Dittany, Call-me-to-you-*
> *Cowslip, Melitot, Rose of the Sun,*
> *Anything green that grew out of the mould*
> *Was an excellent herb to our fathers of old.*
>
> *Wonderful tales had our fathers of old –*
> *Wonderful tales of the herbs and the stars –*
> *The Sun was Lord of the Marigold,*
> *Basil and Rocket belonged to Mars.*
> *Pat as a sum in division it goes –*
> *(Every herb had a planet bespoke) –*
> *Who but Venus should govern the Rose?*
> *Who but Jupiter own the Oak?*
> *Simply and gravely the facts are told*
> *In the wonderful books of our fathers of old.*
>
> *Wonderful little when all is said,*
> *Wonderful little our fathers knew.*
> *Half their remedies cured you dead –*

Four years later, Kipling's generation were desperately grateful for the excellent herbs of their forefathers, when urgently needed drugs ran short during the First World War and the continued supply of medicines in Britain relied on the use of native plants and age-old natural remedies.

Volunteers from all over Britain were organised to gather herbs such as field marigold (*Calendula avensis*) whose antibacterial properties made it a vital medicine in the treatment of wounds and amputations. Marigolds were considered so important that Gertrude Jekyll, the famous garden designer, donated a field at Munstead Wood, her home in Surrey, purely for growing them. Wild garlic bulbs were dug up and used as an antiseptic. The acidity in sphagnum moss had long been known as an antibacterium and vast quantities were gathered from peat bogs and made into wound dressings. Irish moss, a red algae containing carrageenan (a gelatinous substance used in ice cream), was picked all round the coast of Britain and Ireland and used to treat throat damage caused by mustard gas. Shepherd's purse (*Capsella bursa-pastoris*) was used on the battlefield to help control haemorrhaging. The little daisies of the traditional medicinal plant feverfew (*Tanacetum parthenium*) were used to reduce temperatures, and the purple flowers and leaves of marsh or hedgerow woundwort (*Stachys sylvatica*), to staunch bleeding. Comfrey (*Symphytum uplantica*), which contains allantoin, was used in fracture injuries.

The sedative properties of valerian root had been known for centuries and it was now used to treat shell shock and to calm the nerves of Londoners during the Zeppelin bombing raids. As the war ground on and commodities became scarce, the civilian population relied increasingly on wild plants to supplement vegetable and fruit shortages. Towards the end of the war, a crisis occurred when the UK fruit harvest failed and the army ran out of jam. The high command believed jam was a vital antiscorbutic and British Forces consumed 1.5 million pots of it per day. To avert disaster, The Ministry of Food called upon every man, woman and child to strip hedgerows across Britain of blackberries and deliver the fruit to jam-making collection points organised by the War Women's Association.

Twenty-one years after the 'war to end all wars', wild plants were in even greater demand as the German submarine blockades tried to starve Britain into submission during the Second World War. The first emergency to be dealt with was crucial medical supplies, which had virtually run out by late 1940. The Vegetable Drugs Committee was immediately set up, with the task of galvanising the people of Britain to collect and dry medicinal wild herbs. With the assistance of the National Federation of Women's Institutes, the Scottish Women's Rural Institutes and the Women's Voluntary Service for Civil Defence, herb-gathering organisations were set up in every town and village. These were made up of every able-bodied civilian, of school children, Boy Scouts and Girl Guides, who scoured hedgerows, waste ground, canal banks, woods, heaths and moorland for the complex variety of herbs required. To assist herb gatherers, the Ministry of Health, in conjunction with botanical herb merchants such as Brome and Schimmer, provided pamphlets for identification and, more importantly, instructions on how to dry the herbs. In the first year, an amazing 950 tonnes of herbs were collected and delivered to drug manufacturers; an achievement all the more remarkable when one considers that 80 per cent of a plant's weight is lost in drying.

At the end of the war, 3,630 tonnes of the 40 or so different essential herbs were being gathered every year, ranging from those that had been used in the First World War, to skull cap (*Scutellaria galericulata*) for treating spasms and nervous disorders; black horehound (*Ballota nigra*) for the expulsion of intestinal worms; the root of meadowsweet, as a diluent for fevers and agrimony (*Agrimonia eupatoria*) for its anti-inflammatory and blood-staunching properties. Children in the Highlands picked blaeberries (*Vaccinium myrtillus*), which the RAF believed improved the night vision of pilots, whilst in the south, conkers from horse chestnut trees (*Aesculus hippocastanum*) provided the glucose in Lucozade. By 1941, doctors were reporting that British infants were showing symptoms of severe vitamin-C deficiency and an emphasis was placed on gathering rose hips when they ripened in the autumn. Known since long before Culpeper's day for their antiscorbutic properties, hundreds of tons of rose hips were picked and made into a syrup which was provided free to all British children.

In 1939, domesticate food production was only sufficient to support one in three of the population and every year, Britain imported over 70 per cent of consumer commodities, such as meat, cheese, sugar, fruit and cereals. Only a trickle of these got past the Nazi blockades and the Ministry of Food was obliged to impose stringent food rationing within the first months of the war. Suddenly the home front, as the civilian population became known, had to learn to become almost entirely self-sufficient. Food shortage in Britain became even more chronic after the hostilities ceased and the Allies found themselves burdened with providing food for European countries devastated by the fighting. Herb gathering for medical supplies finished in 1946 as stocks of vegetable drugs began to be imported from India and New Zealand, but food rationing lasted another nine, long, grinding years, eventually ending officially in 1954.

It is incredible to reflect, in this age of waste and plenty, that a whole generation had grown up knowing nothing but self-sufficiency and privation. Foraging for wild food, whether along canal banks, hedgerows, waste ground or woodland gave people of all ages and walks of life a deeper understanding of the countryside and its wildlife, than at any time in the preceding century. The nation was, as a result, much closer to and had a greater respect, appreciation and gratitude for their natural environment. After all, at some point during the previous fifteen years everyone, rural or urban, civilian or armed forces had, to a greater or lesser degree depended on nature in a time of need.

I was six years old when rationing ended and I remember my sister and I being taken by our nanny, Nanny Pratt, on daily walks along the lanes and through the fields and woods near my parents' home. On these afternoon excursions Nanny Pratt always took her trug, and whichever wild food was in season at the time was picked and taken back. Wild garlic leaves which carpeted the local woods or mustard garlic from hedge banks, in the early spring; goosefoot, dandelion and chickweed, all full of iron, vitamin B and calcium; the grey-green leaves of common orache, fat hen, shepherd's purse and, later, watercress that grew where a stream broadened as it flowed through a meadow. As nursery food, these were always eaten well boiled, particularly the watercress, which acts as a host to liver fluke. In the summer, when the countryside was a riot of blossom and sweet scent, the trug was filled with flowers for the nursery or wild strawberries and raspberries. As autumn came and the leaves began to turn, we picked blackberries and elderberries for jam; rose hips and sweet haws for the syrup we were given every day and little yellow, rock-hard crab apples for pickling. October was the month for hazelnuts and early November, windblown sweet chestnuts, lying in their spiny green shells among the golds, russets, bronze and ochre of fallen leaves.

We were rarely alone on these trips, as others would often be out gathering wild food at the same time. The gardener's wife, perhaps, picking nettle leaves to make into nettle beer for her husband, or gathering wild hop buds which she made into a pillow to help him sleep when his rheumatism was particularly painful. Sometimes we would see the two elderly brothers who lived in the village picking golden dandelion heads, cowslips, meadowsweet and elderberries to make into wine or carefully picking blackthorn leaves. When dried, these were used as a substitute for tea or smoked in their pipes when tobacco was scarce. On other occasions we would pass places where the broken stems of ground elder and hogweed or fat hen, indicated that someone had recently been harvesting. In the autumn, people bicycled for miles from local towns or came by bus from the cities to pick the hedgerow harvest of blackberries, rose hips or crab apples, returning a month later for the nutting season. For young children of any generation, trailing behind adults as they forage is the beginning of an education in natural history and country lore. They learn to identify what is edible, where to find it and why it grows there, and that some plants might look good to eat but are, in fact, deadly poisonous – such as the black or translucent berries of climbing bryony, or the dark purple ones of deadly nightshade – and never to touch those they do not recognise. At the same time, inquisitive children took an interest in the wildlife around them: the birdsong, darting insects and furtive rustling of unseen little creatures. It was on the outings with Nanny Pratt, as I began to learn about the breeding seasons of animals, that the seeds of my fascination with natural history were sown.

Grey Geese

Thousands of geese stirring out on their shore roosts, as dawn breaks over a storm-swept estuary on a freezing winter's morning, is music of the gods to a wildfowler crouching with his dog among the reeds of a tidal creek. Gulls are always the first to start the dawn chorus on the marsh edge, wheeling and shrieking even before the lights start twinkling in the lonely farmhouse on the other side of the estuary. As the darkness turns to the grey start of a new day, every conceivable species of waterfowl and wader – golden, ringed and grey plover, greenshank and redshank, oystercatcher, bar-tailed godwit, curlew and peewit – start moving along the water's edge, whistling, yapping, tittering or yodelling. As the splinter of light lengthens, herds of ghostly, crescent-shaped curlews fly past, trips of dunlin, flocks of moth-like lapwings and periodically, with a strange vibrating noise, gaggles of pale-fronted brents, flying almost nose to tail in low, wavering lines. Then there is a distant roar as geese lift, an eerie swelling sound growing in volume, followed by the jubilant and ever-increasing 'ang-ang-winking-winking' as skein after skein of pink feet pour up the estuary towards their inland grazing.

The old Labrador starts trembling and whimpering softly as the great armada of geese swing towards them – will this be the moment when all his master's careful studying of moon cycles, tides, flight paths and the weather prove to be correct, and will the wind and cloud cover bring them in range? Maybe. There is nothing contrived, predictable or artificial about 'fowling and this, as much as the mystique of the lonely salt marshes, stark winter beauty of mudflats and foreshore, iodine scent of the sea and the cries of waterfowl, is what attracts people to the sport. To lie unobserved as nature wakes up and starts moving around you is the most wonderful experience and every wildfowler knows, that during the course of a season, there will probably be only a handful of occasions when everything – moon, tides, wind, weather and laborious studying of flight paths, combine to produce the shot one remembers for the rest of one's life.

My love of wildfowling goes back to my childhood, when a period of hard frost put a stop to hunting and there was a flurry of activity as my father packed his wildfowling gear to go north, 'fowling on the Tay: the oiled wool seaman's socks, long johns with buttons at the ankle, vests of real string, leather jerkin, corduroy breeches and the hooded paratrooper's smock with a strap that buckled between the legs. Black rubber hobnailed thigh waders, canvas game bag, compass, otter hunting pole and best of all, the massive, double-barrelled 8 bore and boxes of long, dark-red cartridges. For a small boy, preparations for the trip to Perthshire were almost as exciting as his return and the vivid imagery he created of the snow-covered fields of the Carse of Gowrie running down to the vast reed beds beside the Tay, so tall and dense in places that people had been known to get lost in their depths looking for shot birds. Crouching on the edge of them with Pilot, his black Labrador in the darkness of a freezing early morning, listening to the reeds rustling in the wind and murmuring of geese out on their roosts, as he waited for the dawn. The first glint of daylight reflected on the black surface of the mighty river and the sudden 'whoosh' of wings as geese lifted to fly inland against a lapis lazuli sky. The heart-pumping anticipation of a skein coming in range towards him and the boom and kick of the big gun.

I vividly remember much later, just after I left school, sharing a section of broken stone wall with him on the Northumbrian coast one warm, windless, early October afternoon, as we waited for the tide flight.

We could see a big raft of wigeon out on the bay and occasionally, trips of teal zipped along the water's edge. It was very peaceful watching eider duck and white-fronted brents fly by, knots and dunlin performing their synchronised aerobatics, and listening to the silly laugh of shelduck or hoarse cackle of mallard. It would be some time before the tide pushed birds closer and the wigeon started moving about, and puffing contentedly on his pipe, my father began reminiscing about the history of wildfowling. Taking me back to a time when a quarter of Britain and most of East Anglia was wetlands, teeming with every conceivable variety of waterfowl that migrated south in September to escape the winter in Iceland, Greenland and northern Europe, returning at the approach of spring.

For centuries, marsh dwellers supplied inland communities with reeds for thatching and flooring, salt for winter survival, baskets of eels, pike and perch, and wildfowl, netted as they flighted in to marsh ponds. Reclamation of the marshes by the Romans, Normans and through successive monarchies during the Middle Ages, made little impact on the vast acreage of wetlands, or the wildlife population. The first real pressure came in 1626, when Charles I commissioned the Dutch engineer, Cornelius Vermuyden, to drain 40,000 acres of the Isle of Axholme in Lincolnshire, to the fury of the local marsh dwellers. This was nothing compared to the 750,000 acres of fens drained in Norfolk, Lincolnshire and Cambridgeshire between 1632 and 1657, despite the determined efforts of the 'Fen Tigers', who sabotaged the engineering works whenever they could.

As more wetlands and wildfowl habitat was reclaimed during the agricultural revolution, the professional 'fowlers, now known as 'market gunners', dressed in sealskin caps and leather waders, stinking of the goose fat they smeared over themselves as a protection against the bitter cold, struggled to make a living with primitive shotguns, using their unique knowledge of the behaviour of marsh bird life. With a rapidly spreading urban population and ever-increasing demands for more birds, the market gunners were driven to risking their lives by stalking duck sitting on open water, or exposed mudbanks with big bore guns – effectively a small cannon – mounted on flimsy 'gunning punts' in the hope of making a big bag with one shot. The skill involved in approaching a 'sitting' across sea water in daylight and unpredictable weather, lying prone in a canoe-shaped boat with a cannon mounted in the bow and a freeboard of only inches, attracted the attention of the celebrated sportsman, diarist and author, Colonel Peter Hawker (1786–1853). Hawker's sporting diaries from 1802 to 1853 were avidly read and he was followed by a host of Victorian and Edwardian gentleman gunners, naturalists, authors and artists, such as Lewis Clements, John Guille Millais, Sir Ralph Payne-Gallwey, Archibald Thorburn, Frank Southgate and Abel Chapman. They were only a few of those who discovered kindred spirits among the longshore 'fowlers and puntsmen, finding the inspiration to paint or write from the irresistible lure of remote salt marshes, the haunting cry of waterfowl, murmuring of geese out on their roosts and the distant rubble of surf.

As the railway network expanded in the nineteenth century, these isolated areas became easily accessible and the rights which permitted anyone to shoot on the Crown foreshore between high and low tide, increasingly abused. Everything that flew was quarry in those days and continuing wildfowl habitat erosion through land reclamation, the Victorian craze for egg collecting and taxidermy, commercial duck decoys – one on the Ouse near St Ives was sending 3,000 brace a week to London's Leadenhall Market – had already caused several rare geese and waterfowl species to become extinct. By the end of the century, professional wildfowlers and experienced devotees had become desperately concerned at the lack of control and indiscriminate shooting on the foreshore.

One of these, Stanley Duncan, an engineer, naturalist and experienced wildfowler who 'fowled on the marshes round Sunk Island on the Humber Estuary, founded the Wildfowlers Association of Great Britain and Ireland in 1908, with the support of Sir Ralph Payne-Gallwey, the association's first president. Their aim was to encourage wildfowlers to form themselves into clubs and acquire the shooting rights on Crown foreshore and the conterminous marshes, bringing them under the control of responsible people whose love of wildlife would ensure their conservation, whilst enabling a traditional pot-hunting pastime to continue. From small beginnings, WAGBI grew into an organisation which set standards for wetland conservation in Europe, America, Canada, Australia and New Zealand. In 1981, WAGBI expanded to encompass all shooting and conservation issues, changing its name to the British Association of Shooting and Conservation.

There are no professional wildfowlers left, the sale of wild geese is illegal and wildfowling clubs impose bag limits on the number of birds that may be shot on any day during the season (1 September – 31 January inland, below the high-water mark on the foreshore 1 September – 20 February). Under the Wildlife and Countryside Act 1981, legal wildfowl quarry is restricted to Canada, greylag and pink-footed geese, with white-fronted geese only in England and Wales; common pochard, gadwall, goldeneye, mallard, pintail, shoveler, teal, wigeon and tufted duck. Some, such as teal, wigeon and young pink-feet geese – which generally fly at the back of a skein – are better eating than greylag or pochard, but each have gastronomic qualities of their own and none find their way into a bag without a great deal of hard work and dedication.

The efforts of WAGBI and BASC have preserved an historic part of our sporting heritage. Through them, I and many others, have game books of treasured memories of salt marshes, foreshore and mudflats; butterfly weather and bitter, howling wind; failure and occasional success. More importantly, they have enabled history to repeat itself in the days I have spent passing it on to the next generation. Kneeling behind a hedge on the side of Wardlaw as skein after skein of pinks flighted up the Nith Estuary towards us; being covered in snow as we crouched in the reeds beside the River Kent where it flows into Morecambe Bay, waiting for the dawn; looking across to Holy Island and listening to wigeon whistling as the tide crept in; watching with fatherly anxiety the progress of a punt inching its way down the Tay. Or simply standing together in the farmyard on a moonlit November night, listening to the Gabriel hounds calling to each other as they flight inland to feed.

The Language of Field Sports

At the Boxing Day meet last season, I overheard a small boy on a leading rein say to his mother, 'Why is that hound missing half its tail?' To be immediately told, as countless generations of children have before him, that a dog has a tail, but a hound has a stern and remembering the difference is terribly important. The gravitas of this introduction to the language of field sports was obviously not wasted on our budding Nimrod, for as we moved off, I could see his lips moving and his brow furrowed in concentration.

Knowledge of the complex terminology of every aspect of the chase has been the essential education of young gentlemen since antiquity, when hunting for survival developed into the true sport of kings and princes. The pharaohs of ancient Egypt invented field sports, passionately hunting everything from hippos and elephants to gazelles and wild bulls. It was the courtly pleasure of the kings of Assyria, who considered hunting proper training for war and hunted lions from their chariots in a demonstration of royal courage, whilst the Persians hunted stags and wild boar on horseback. The Greeks were great hound men, breeding deep scenting hounds for hunting boar and stag, a type of beagle for hare hunting and fast sighthounds for coursing. Xenophon the historian, who believed hunting was God-given and a necessary part of a well-rounded young man's education, wrote *Cynegeticus,* a detailed treatise on hunting for their benefit in 394 BC.

The Romans were no less devoted to the chase, importing different hounds bred by the Celts in Britain long before the Invasion: alaunts, rough-coated sighthounds similar to deerhounds, used hunting for boar, wolves and deer. Light-framed vertagus, the ancestor of the greyhound, bred for coursing hare, and scenting hounds such as the segusian and the smaller agassian, described by Oppian, as 'a strong breed of hunting dog, small in size but no less worthy of great praise.' The Romano-British bred fine horses and hunted boar, wolves, wild cattle, fox, red and roe deer, with wealthier landowners creating game parks stocked with fallow deer and brown hares. Coursing was raised to a new science where the purpose was not to kill the hare, but to provide a challenge between the fastest and most agile of all creatures against the speed of two matched vertagri. The ethos, which remained unchanged until match coursing was banned by the Labour Party in 2004, was described by Arrian as: 'the aim of the true sportsman with hounds is not to take the hare, but to engage her in a racing contest, or duel, and he is pleased if she escapes.'

Almost every advance in civilisation introduced by the Romans was reversed by the Barbarian hordes that swept across Europe and invaded Britain during the Dark Ages. The finesse of hunting was a thing of the past and it was the fiercer game, boar and wolves which provided all the drama and violence that appealed to those gross, illiterate tribal kings and gave young warriors an opportunity to prove their bravery. The ancient traditions of hunting were reborn under the Frankish kings and by Charlemagne in particular, who created Royal Forests across his enormous empire in which chosen nobles were granted hunting privileges. Charlemagne made hunting into a major state institution, introducing pomp, ceremony and ritual, and it became once again, an essential part of every young nobleman's education, combining knowledge of natural history with physical courage, strenuous activity, good manners and gentlemanly respect for the quarry.

Frankish hunting practices were introduced to England by King Alfred, who was a great admirer of Charlemagne and had spent part of his childhood at the court of his grandson, Charles the Bald. Alfred was a dedicated hunting man, who was 'a most expert and active hunter and excelled in all branches of that most noble art,' bringing up his children to 'practice the human arts, namely hunting and horsemanship, and those other pursuits that befit noblemen.' He re-introduced the sophisticated science of coursing with greyhounds, created a Horse Thane charged with the task of improving the quality of horses in England and established Royal hunting lodges across his kingdom, appointing foresters and verderers to manage the deer herds. It was during Alfred's reign that we read of the practice of 'leasing', a flogging with a set of hand-couples dispensed on the spot to any young noble who arrived late at a meet, hollered a wrong deer, left before the kill, or made a mistake with any of the hunting terms.

The Anglo-Saxons believed that although the laws of trespass should be observed, wild animals were *res nullius;* belonging to no one. Anyone had the right to hunt on their own land and if the quarry was hunted on to someone else's and killed there, the venison was shared between both. The land reserved for the monarch's exclusive use was hardly disproportionate and the penalties for poaching, not unduly harsh, but this relaxed and tolerant attitude to game was to change dramatically with the Norman Conquest. Britain now became subject to European feudal law and under feudal law, land belonged by divine right to the king and the same divine right of ownership applied to all game species protected by ferocious forest laws. Duke William doled out hunting rights to his nobles and the Church and these rights, like the game in them were strictly categorised in order of precedence. The noble beasts of forest or of venery: the hart, the hind, the hare, the boar and the wolf. The beasts of chase were the buck, the doe, the fox, the marten and the roe. The beasts and fowls of warren were the rabbit, the pheasant, the partridge, the woodcock and later, rail, quail, bustard and waterfowl.

England now became an integral part of France and largely indistinguishable from the rest of it. French hunting methods were adopted and for the next 300 years, whilst it remained the language of the court, the increasingly elaborate and complicated language of field sports was in French. The crusades of the twelfth century led to the Age of Chivalry and hunting became seen as a worthy, semi-religious, knightly occupation, blessed by the Church. For two centuries, from about 1150 to 1350, the education of a gentleman's son was as a squire in the household of a noble, where he learnt courtly manners and every aspect of field sports: each method in which the hunting of different game was conducted, each year in the development of the quarry species and each of its body parts. Every stage of the chase, the different notes of the horn, each feature of a hound's behaviour, the complex ritual of the kill, the 'breaking up' or 'unmaking' of the carcase and the curée – the offal given as a reward to the hounds. Every single species of game, whether fur, fish or fowl had to be carved according to rigid individual specifications based on their standing in the codes of hunting. Such was the emphasis on the exact procedure of carving, that the golden spurs of knighthood could not be granted until the correct method and terminology had been painstakingly learnt. The convoluted and stylised rituals of hunting were expressions of respect for the hunted quarry and knowledge of the language of field sports was the badge of a gentleman. To make an error in the Terms of the Art of Venery or the elaborate and whimsical dictionary of hawking, was to commit an unforgivable and damning social solecism.

The earliest descriptions of the style of early medieval hunting are in treatises written in Norman French; *Le Art de Venerie* (1323) dictated by William Twici, huntsman to Edward II. *Le Livre du Roy* (1328) by Henri de Ferrieres and probably the most famous work on hunting of all time, *Livre de Chasse*, written by Count Gaston de Foix-Béarn in 1387. With painstaking detail, the count records the different stages of hunting the various quarry species and their behaviour, offers advice to less well-heeled gentry on how to run a sporting establishment without bankrupting themselves and is even sympathetic towards the peasant poacher, because he too is a sportsman at heart. The first work on the chase in the English language was *Master of Game*, written by Edward, 2nd Duke of York in the early fifteenth century and whilst much of it is an exact translation of *Livre de Chasse*, Edward included five detailed chapters on the different methods of hunting in England, which provides the major source of our knowledge of the minutiae of English hunting at that time.

The social structure of England underwent dramatic changes after the Black Death effectively put an end to feudalism. The labour shortage created possibilities for able peasants to amass enough money to acquire land and with fortunes being made from wool, their families were able to become gentrified within a couple of generations. In 1390, the forest laws protecting the rights of the king were replaced with laws protecting the rights of the gentry and those entitled to hunt were now qualified simply by income. This created the unique and abiding characteristic of English field sports; there was no class distinction. The yeoman was entitled to hunt as freely as his armigerous neighbour and there was nothing to stop a small tenant farmer or tradesman from following the hounds of the gentry on equal terms with all but the Master of Hounds.

Becoming gentrified was conditional on learning the detailed law, language and rituals of hunting and in 1486 a series of essays were published specifically for the benefit of the newly landed and socially ambitious. *The Boke of St Albans* – unconvincingly attributed to a Dame Juliana Berners of whom nothing can be verified – are manuals covering the customs and terminology of heraldry, hunting and hawking, about which every man claiming to be gentle was expected to be familiar. The treatise on hunting ends with a list of the correct company terms of 'beestys and fowlys', still in use today, such as: a *bury* of rabbits, a *business* of ferrets, a *fall* of woodcock, a *walk* of snipe, a *sounder* of boar and so on. *The Boke of St Albans* quickly acquired a further essay on angling and was in such demand that there were twenty-two reprints between 1486 and 1616 – more than any other work of literature over the same period, except the Bible – and was frequently plagiarised by others, including Gervase Markham who used much of the material in his *Gentleman's Academic* of 1595 and Izaak Walton, in his *Compleat Angler* of 1653.

Archaic Norman French is still the hound language of hunting in *The Boke of St Albans,* but by the end of the seventeenth century, hunting in England had become remote from the elaborate protocol of France, both in terminology and method, a distance which would escalate throughout the eighteenth and nineteenth centuries until only the cry of 'il est haut, il est haut' has remained as *'tally-ho'*. Our 'modern' language of field sports has evolved into something equally as complicated and idiosyncratic as that of previous centuries; thus, a dog has a tail, a hound a stern, an otter a rudder, a deer a single, a fox a brush and a hare a scut. We count the bag for snipe, duck and rabbits in couples and halves, grouse and partridge in brace and halves, but not pheasants; we call 'mark cock' when we see woodcock, but 'cock forrard' for a cock pheasant; a salmon is awarded the honour of being simply a fish, whilst lesser species are named specifically. These are only a minute fraction of the proper terms in the complex language of field sports which a true sportsman has a duty to learn out of respect for the quarry and to preserve our sporting heritage for future generations.

Campanology

Of all the symbols of the twelve days of Christmas, the holly, ivy and mistletoe; the wreathes, tree and decorations; the turkey and plum pud, none is so synonymous with the spirit of goodwill, peace and joy, or more ancient, than the sound of church bells, summoning the faithful to celebrate the birth of Christ at midnight Mass on Christmas Eve and to Matins on Christmas Day. The most symbolic and evocative of them all, is the eerie peal of half-muffled bells ringing out the old year, ending with the great tenor bell solemnly tolling twelve times as the hour reaches midnight. There is a pause whilst the muffles are removed; then all the bells start again, a joyous, exuberant, cacophony welcoming the New Year, a glorious sound, full of power, hope and promise:

> *Ring out the old, ring in the new,*
> *Ring, happy bells, across the snow:*
> *The year is going, let him go;*
> *Ring out the false, ring in the true.*
> *Tennyson, 'In Memoriam', 1850*

The sound of church bells has been deeply rooted in the culture and heritage of Britain for nearly 1,500 years, and almost everyone lives within hearing range of them. The association between bells and the church has its origin in the pagan belief that the sound of a bell ringing drove away evil and like many pagan beliefs, the early Church conveniently converted it into a Christian practice. Handbells were used by early missionaries to summon people to worship and tolling bells were reputedly introduced to churches in AD 400, by Paulinus, the bishop of Nola in Campania and the custom gradually spread across Europe, arriving in Britain towards the end of the sixth century. The abbeys of Wearmouth and Whitby are recorded as having bells in AD 680 – the Venerable Bede mentions the bell tolling for the death of St Hilda, the founding abbess of Whitby Abbey, was heard 14 miles away at the monastery of Hackness. Ecgbert, Archbishop of York, issued instruction in AD 750 that bells were to be rung for the canonical hours of devotion and St Dunstan, Archbishop of Canterbury, decreed that all churches in his diocese should have them.

Bells are made of bronze and by the Middle Ages, every church had at least one and the wealthier cathedrals, abbeys and monasteries, as many as six. Some of these were enormous – in 1316, Prior Henry of Eastry gifted a bell weighing 4 tons to Canterbury Cathedral, dedicated to St Thomas à Becket. Many bells were cast on site by itinerant artisans with a basic knowledge of smelting and mould making, travelling from church to church; others were craftsmen who set up permanent foundries. Peter de Weston and his successor, William Revell, were casting bells in their Aldgate foundry in the mid-1300s, but there were many others in towns and cities such as Durham, Salisbury, Gloucester, Reading, Norwich, Colchester, or Wokingham, attracting business from the surrounding countryside.

Early bells were hung from an axle or spindle to which a rope was attached, enabling the bell to be tolled and where a church had multiple bells, these were limited to being rung in succession, producing only a fairly monotonous, discordant sound for a lot of hard work. The fifteenth and sixteenth centuries were a period of experimentation to make bells more controllable and mellifluous. Initially a quarter-wheel, then a half-wheel and finally whole wheel were attached to the spindle with the rope passed round a groove in the wheel, which made it possible for a bell to be swung through 360 degrees. As the bell reached the top of the pendulum, it could be paused momentarily and the time of its ringing controlled. With sufficient control over the timing, a 'ring' of bells could be rung up and down the scale, from the lightest to the heaviest and vice versa, a bell-ringing practice known as 'rounds'.

Bell-ringing underwent a hiatus during the Dissolution of the Monasteries and Reformation, and in the reign of Elizabeth I only one bell was allowed to be rung as a summons to worship. Sundays were seen as a day of rest and entertainment, when all manner of sport was permitted among them, bell-ringing for pleasure and this now became a secular pastime, with very little connection to the building in which the bells were hung. Bell-ringers were paid to ring for events of regional or national importance – in 1586, the bell-ringers of St Margaret's, Westminster, were paid a shilling each to ring at the beheading of Mary, Queen of Scots. At the beginning of the seventeenth century, bell-ringing began to become standardised by the formation of bell-ringing companies, with rules similar to the London livery companies. Two of the oldest surviving companies are The Company of Ringers of the Blessed Virgin Mary of Lincoln, formalised in 1612, whose members ring the bells in Lincoln Cathedral today and the Ancient Society of College Youths. Established in 1637 and based in the City of London, they provide ringers at important events in St Paul's Cathedral and Westminster Abbey.

By now, bell-ringers had discovered that by changing the sequence in which bells were rung from the treble to the tenor, a distinctly musical effect was achieved and this method of ringing became increasingly popular. An important milestone in the early days of campanology and 'change' ringing as a science, was the publication in 1667 of *Tintinnalogia,* by Fabian Stedman and Richard Duckworth, which detailed 'plain and easie Rules for Ringing all sorts of Plain

Changes.' Stedman used mathematics to work out how bells of different weights could be rung in a series of permutations without any of the changes being repeated; for example, 720 continuous changes were feasible with a tower of 6 bells. The success of *Tintinnalogia* encouraged Stedman to write *Campanalogia* in 1677, which contained fifty-three new ringing methods using five, six, seven or eight bells, many of which are as popular today, as they were in the seventeenth century. To Stedman is owed the complex system which makes up a magnificent peal of bells, a 'performance' or 'extent' of 5,040 changes taking over three hours of uninterrupted ringing, where the bells are never sounded in the same order twice. The first peal of bells is believed to have been rung by the Society of College Youths in the church of St Sepulchre-without-Newgate, Holborn, in January 1690, but the first documented peal was rung on seven bells in St Peter Mancroft, Norwich, in May 1715. The more bells, the greater the number of permutations – in 1756, to celebrate the birthday of the Duke of Cumberland, a continuous peal of 1,040 changes lasting six and a half hours, was rung on ten bells in the church of St Laurence, Reading.

To the outrage of reputable bell-ringing societies, during the eighteenth and early nineteenth centuries, bell-ringers, particularly in rural areas, gained the reputation for drunkenness and obstreperous behaviour, swearing, smoking, gambling and drinking in the belfry, and refusing to ring unless they felt like it. In 1832, the bell-ringers at All Saints in High Wycombe rang to celebrate the passing of the Reform Bill, but refused to ring a few days later when the Bishop of Oxford paid his annual visit because they suspected him of voting against the bill. The relationship between the clergy and ringers became so bad that some incumbents tried locking bell-ringers out of the belfries, with the ringers responding by breaking in to what they considered their own personal domain.

The Victorian reforms of the Church of England enabled the clergy to take back control of the belfries and bell-ringing became re-aligned to Christian practices, with bell-ringers encouraged to be part of the congregation and appoint tower captains responsible for the attendance and general conduct of the ringers. Between 1851 and 1875, 2,438 churches were built or rebuilt and many bell towers refurbished, with bells recast and new bells added by foundries such as Taylor's of Loughborough, Watson of Newcastle, Bush of Bristol, the Albion Foundry in Leeds or Mears of Aldgate.

Church of England reforms drove a resurgence of interest in bell-ringing from all walks of life, with societies and guilds being formed all over the country. In 1891, Sir Arthur Heywood, the 'Bell-ringing Baronet', founded the Central Council of Church Bellringers, an advisory body, which today acts as a coordinating organisation for education, publicity and codifying change ringing rules worldwide. They provide advice on maintaining and restoring full circle bells, and publish *Ringing World,* a weekly journal which records notable ringing performances, carries features on bells, change ringing, bell towers and ringers. By the late 1800s, women began to take up bell-ringing, becoming part of what had previously been an all-male preserve and in February 1896, at the age of nineteen, the redoubtable Miss Alice White became the first woman to ring a full peal as part of a team of eight ringers in St Michael's Church, Basingstoke. Such was the increasing popularity of change ringing among women, that a Ladies' Guild of Change Ringers was formed in 1912, the same year an all-women team of eight ringers rang a full peal at Christ Church, Cubitt Town on the Isle of Dogs.

Bell-ringing continued to go from strength to strength through the 1900s and in 1997 a Lottery grant of £3 million went towards 150 separate bell restoration projects and a recruitment drive which attracted 5,000 new ringers, who learnt to ring in time for Ringing in the Millennium. This was the largest national ringing event ever staged, when 95 per cent of UK church bells rang together at twelve noon on 1 January 2000. To signify the long heritage of bell-ringing, in 2012 a floating belfry with eight bells cast for the occasion by the Whitechapel Bell Foundry, led the Queen's Diamond Jubilee Thames Pageant with a joyous peal. In 2017, in the run-up to the nationwide events marking the centenary of the end of the First World War, a new campaign was launched to commemorate the 1,400 bell-ringers who died in the war by recruiting the same number of new ones, which exceeded expectations by attracting over 1,700.

There are over 5,000 bell towers in Britain and around 40,000 active bell-ringers of all ages and apart from the Ladies' Guild, sixty-three bell-ringing guilds, societies, or associations are affiliated to the Central Council. Bell-ringing is social, fun and good exercise, both physically and mentally. Church congregations may be dwindling, but the bells will always ring out on New Year's Eve.

Sighthounds

Sighthounds or gazehounds – those that hunt entirely by sight and speed, rather than scent, are among the most beautiful and graceful of dogs, with a lineage dating back at least 5,000 years, and in all probability, far longer. Sighthounds are believed to be descended from the primitive running dogs which hunted across the vast, low grasslands of the Fertile Crescent of the Middle East. By selective breeding, the ancient Egyptians are credited with developing an agile, long-legged sighthound, with massively powerful hindquarters, fast enough to course gazelle, antelope, hare and ostrich. Sighthounds spread radially across Eurasia and into North Africa as early civilisations came and went, and with each, were the treasured possessions of monarchs and nobles, who celebrated them in statutory mosaics and murals, indeed, no animal in history, apart from the horse has been more represented in art.

With the passage of time, true sighthounds developed into distinct geographical breeds: the bare-skinned desert hounds of the Middle East, North Africa and India; the heavy-coated sighthounds of Afghanistan and through Russia. Across Europe, the Celts, who were great hound men and particularly fond of coursing hares, bred a particularly fine greyhound type much admired by the Romans, called the Celtic hound, or the vertagus. In 2019, Bonham's sold two marble statues of Celtic hounds from the second century AD, which although 2,000 years old, are indistinguishable from the Italian greyhound of today. The great Celtic tribal chiefs in Britain kept extensive kennels of sighthounds bred specifically for different quarry species: light vertagri for match coursing hares and the taller, more robust, rough-coated hounds of

Celtic myth and legend, for hunting deer, boar and wolf, arguably the ancestors of our modern greyhound, Scottish deerhound and Irish wolfhound.

Since the Hunting Act 2004, the number of sighthounds in Britain has declined, but there are still enough of these magnificent aristocrats to be seen at game fairs, gliding insouciantly through the commonality of other canines, with their flowing, springy gait; elegant, aloof and assured of their own superiority.

Scottish deerhound

Until the early nineteenth century, when the breed became standardised, these tall, strong-boned and rough-coated sighthounds, were known as Highland greyhounds or Scotch staghounds. As a breed they are of immense antiquity and were for many centuries, the treasured companions of Scottish monarchs and clan chieftains. Specifically bred for coursing red deer and although not as fast as greyhounds, they had the stamina to follow a running deer over rocky ground, with the strength to hold a 300-pound stag. They coursed in couples, with one, the 'high dog', trained to run wide round the corrie tops to stop a stag crossing the skyline and turn it, whilst the 'low' dog ran straight in after the beast.

By 1810, sheep replaced deer on many Highland estates and the Highland greyhound fell on hard times, with only a handful being bred true, by enthusiasts such as the McNeills of Colonsay, the Dukes of Gordon and Glengarry MacDonnells. There was a resurgence of interest in the breed, driven by the romantic Scottish novels of Sir Walter Scott, who owned a deerhound called Maida, which he described as 'a most perfect picture of heaven'. Queen Victoria had another called Hector, one of the many painted by Sir Edwin Landseer.

The Scottish Deerhound Club was founded in 1891 and registered with the Kennel Club in 1901, with minimum desirable height of 30 inches for dogs and 28 inches for bitches. That the breed survived through the twentieth century is due to a handful of breeders, among them Marjorie Bell, Norah Hartley, Agnes Linton and Anastasia Noble. Poaching gangs persecuting red deer with lurchers led to a ban on hunting deer in Scotland and, from 1954, coursing meetings under the National Coursing Club rules were held on mountain hares, in an attempt to preserve the hunting instinct. Since 2004, these are illegal and deerhounds have become simply family pets and exhibition dogs.

Irish wolfhounds

These huge, amiable, shaggy hounds are close cousins to the Scottish deerhound and have a Kennel Club breed standard of a minimum of 32 inches at the withers for dogs and 28 inches for bitches. Bred by the Gaels for hunting boar and wolves, with a reputation for ferocity when aroused, their popularity spread across Europe and they continued to be used for controlling wolves in Ireland well into the seventeenth century. The last wolf in Ireland was killed in 1786 and by then, very few wolfhounds were left, with the breed assumed to be extinct by 1836.

In 1863, Captain G. A. Graham of Dursley, Gloucestershire, decided to recreate the Irish wolfhound, using the biggest and best available examples of deerhound and Great Dane, breeds he believed to be descended from wolfhounds. Borzoi and rough-coated Tibetan mastiff were later used as outcrosses and by 1885, Graham's wolfhound was breeding true, the Irish Wolfhound Club was formed, the breed standard agreed and in 1886, it was accepted by the Kennel Club. In 1902, he presented a wolfhound to the newly formed Irish Guards and a wolfhound remains the regimental mascot today.

Borzoi

The magnificent Russian wolfhound, with its deep chest, long head, arched back and silky coat, developed in the seventeenth century from a mixture of breeds, among them Saluki, Polish greyhound, Scottish deerhound and the Russian Loshaya, to produce a tall, graceful sighthound which became favourites of the Imperial Family. Many, such as Grand Duke Nicholas kept enormous kennels; up until the Revolution, he had two packs of 120 foxhound-type scent hounds, 150 borzois and 15 English greyhounds at his Perchino estate. Hunts on a massive scale lasted several days with armies of beaters driving hares, foxes or wolves onto open ground to be coursed. Wolves were the optimum quarry and when sighted, a brace of borzois would be slipped, trained to seize the wolf by either side of the neck and hold it until a mounted huntsman arrived. Apart from the sport, these hunts were a test of prowess and only the fastest, bravest and most intelligent borzois were kept for breeding.

Queen Victoria was the first person to own a borzoi in Britain, when she received one as a gift from Tsar Nicholas I in 1847. Interest in these spectacular hounds spread here and across Europe, with the Duchess of Newcastle establishing her famous Nottinghamshire kennel and founding the Borzoi Club in March 1892. Borzois were registered with the Kennel Club in May of the same year and the breed standard set at a minimum height of 29 inches at the withers for dogs and 27 inches for bitches.

The great kennels of the Imperial Family were destroyed in the orgy of destruction during the Russian Revolution, but by then, the best bloodlines were safe in the West.

Saluki

These dignified, striking sighthounds of the Middle East, with narrow bodies, silky coats and feathering on the ears and hindquarters, appear almost fragile on their long, delicate legs and disproportionately large pads. They are capable of achieving speeds of over 40 mph and have remarkable stamina for long distances over rough ground, their heavily padded feet absorbing the impact. They have an incredibly ancient pedigree dating back to the pharaohs. Later, they were highly prized by the sheiks of the nomadic tribes of the Middle East for hunting, hare, fox and gazelle, in conjunction with a falcon trained to distract the quarry by attacking its head.

Salukis were rare in Britain until the end of the nineteenth century, when the Honourable Florence Amherst started her kennel at Foulden Hall, near Thetford, Norfolk. Florence had seen salukis in the Nile Delta when accompanying her father, the Egyptologist Lord Amherst, and acquired a pair in 1895 from the kennel of Prince Abdullah of Transjordan. There was an influx of salukis after the First World War, brought back by officers returning from the Middle East, one of whom, Brigadier-General F. Lance joined Florence Amherst in forming the Saluki or Gazelle Hound Club in 1923, with the breed and standard accepted by Kennel Club at 23 to 28 inches at the shoulder for dogs, and proportionately smaller for bitches.

Whippet

There had been different sizes of greyhounds for hundreds of years before Edward of Norwich, Duke of York described the advantages of small, medium and large greyhounds for different quarry in *Master of Game*, his treatise on medieval hunting, written between 1406 and 1413. Standing around 20 inches at the withers, whippets fall between the little Italian greyhounds at 15 inches and a true greyhound at between 25 and 29 inches.

As anyone who has owned one knows, whippets are superb ratting and rabbiting dogs, but really became popular during the nineteenth and mid-twentieth centuries at walked-up competition hare-coursing meetings and for organised racing. They were known as the 'poor man's racehorse' and whippet racing to the 'rag' – where the dogs raced towards their owner who was waving a cloth – was phenomenally popular, particularly in the north. The breed became recognised by the Kennel Club in 1891 and despite the decline in whippet racing, these enchanting, sporting little dogs remain popular.

Greyhound

Thousands of years of selective breeding went into creating a dog fast enough to match the speed of a brown hare, the fastest European land mammal, capable of reaching a speed of 45 mph. Owning such a sighthound was the ultimate status symbol and match coursing, the challenge between a brace of sighthounds and a hare, was the obvious progression. From the very beginning, the object was never to kill the hare, an ethos already ancient when the historian Arrian wrote in the second century AD: 'For coursers, such are at least true sportsman, do not take their dogs out for the sake of catching a hare, but for the contest and sport of coursing, and are glad if the hare meet with an escape.' Greyhounds were so highly regarded during the Middle Ages, that Edward III adopted the image of a greyhound for his great seal and subsequently, a white greyhound became one of the heraldic supporters in the Royal coat of arms of Tudor kings.

Popularity of coursing grew exponentially, to the extent that Queen Elizabeth I instructed her Earl Marshal, the Duke of Norfolk, to produce a formal code of conduct. The eighteeen 'Laws of the Leash', established the distance of the head start a hare was given before the brace of greyhounds were slipped and a complicated method of scoring points for each time the hare was turned by either dog. As ever, the object was not to kill, but a test of skill and agility.

In 1776, the Earl of Orford, created the 'perfect' greyhound, from which the modern breed descends by crossing greyhounds with deerhounds, Italian greyhounds and even bulldogs. The nineteenth century was the golden age of coursing and greyhound breeding, with coursing clubs springing up all over the country. The emergent rail network fuelled meeting attendance and in the Victorian heyday, over 10,000 flocked to the bare marshes north of Liverpool for the annual three-day Waterloo Cup. Such was the national interest, that in 1871, the winning greyhound was summoned to Windsor Castle by Royal Command, to be presented to Queen Victoria.

In 1858 the National Coursing Club was formed, which controlled coursing in the same way the Jockey Club controls racing and, in 1882, the NCC created the original Greyhound Stud Book. The intense competition of Victorian coursing produced a remarkable creature with classic looks, incredible speed, stamina and courage and all dogs running on the coursing field and later, on the track, had to be registered in the Stud Book. The lineage of greyhounds running on track and field in Britain, Ireland, America and Australia today, is directly traceable to past winners of the Waterloo Cup.

The last Waterloo Cup was run in February 2005 and it is a paradox that the Hunting Act 2004, that monument to ignorance and prejudice, not only destroyed centuries of culture and sporting heritage, but stopped the one field sport where the object was not to kill the quarry.

Yule Log

In the hall of the rambling Elizabethan farmhouse of my childhood home, was a wide, inglenook fireplace and every Christmas Eve, the gardener dragged in an enormous Yule log and with much heaving and grunting, balanced it across the fire dogs. This would be lit by the remaining piece of the previous year's log and the fire had to burn for the twelve days of Christmas. A roaring log fire when outside all is in the grip of bleak midwinter and the wind thunders in the chimney, is as much a part of Christmas Eve as the tree, the decorations, the holly, ivy and mistletoe and as with so many of our Christmas traditions, its origins pre-date Christianity by at least a millennium.

The winter solstice, the shortest day and the longest night, occurs around 21 December and for a period of about a fortnight, our Neolithic ancestors and the Iron Age Celts lit huge bonfires to conquer the darkness, and held sacrifices in a desperate plea for the sun to be reborn, bringing its promise of light, warmth, regrowth and fecundity. The tradition of burning a Yule log came to these islands after the collapse of the Roman Empire and the influx of tribes from Scandinavia. An entire tree trunk of oak or ash with the branches lopped off was dragged into the great halls of tribal Saxon chiefs and one end placed in the open fires. The halls were decorated with evergreens, holly, ivy and mistletoe, a custom inherited from the Roman midwinter festival of Saturnalia and during the twelve days of feasting to celebrate the return of the sun, the trunk was gradually fed into the fire so that it burnt continuously. Endless superstitions were attached to it: the sparks spiralling upwards were believed to represent the successful birth of different species of livestock; the ash protected the house from natural disasters and, if mixed with water, was a general cure-all for both humans and animals. Difficulty in lighting the Yule log was considered ominous; if the flames cast a person's shadow without the head, it presaged their death in the coming year and it was a complete calamity for everyone if the fire went out.

We have the Church to thank for preserving the ancient pagan seasonal festivals that are now part of the Christian calendar – Christmas, Easter, May Day, St Valentine's Day and harvest festivals, to name only a few. The main thrust of Christianity in Britain started in AD 595, when Pope Gregory sent a mission of forty monks led by Augustine, prior of St Anthony's Abbey in Rome and later the first Archbishop of Canterbury, from Rome to England, with instructions to convert the heathen inhabitants. Augustine was advised that the only hope of success was to allow the outward form of the old pagan ceremonies and beliefs to remain intact, but wherever possible to superimpose Christian ceremonies on them. Pope Gregory's mandate of conversion through coercion was brilliant in its simplicity and he surmised, accurately, that the easy-going Saxons would not object if the seasonal festivals of the heathen calendar were Christianised, as long as no one stopped them celebrating. Gradually, over several centuries, the principal pagan feasts became days honouring the life of Christ or one of the saints, but the most important of them all, the great winter festival of fire and light, of new life and lengthening days, was so deeply enshrined in the hearts of the population, that the Church adopted it as the pre-eminent day in their own calendar.

Except in the great country houses, with their huge hungry fireplaces big enough to accommodate an enormous chunk of wood capable of burning for the twelve days of Christmas, hearths became smaller, the Yule log was cut to fit and only required to stay alight overnight. With the population increasingly urbanised after the Industrial Revolution and fireplaces yet smaller, wood was replaced by coal and the Yule log simply became a table centrepiece surrounded by candles and decorated with holly and ivy. In the late nineteenth century, the ever-popular Yule log cake or Bûche de Noël began to be sold by confectioners. The invention of this chocolate Swiss roll made of genoise sponge filled with buttercream and decorated to resemble a Yule log, is attributed and Pierre Lacam, at one time pastry chef to Charles III of Monaco.

Of all the berry-bearing winter evergreens we use to decorate our homes at Christmas, which the ancients ardently hoped were assurances from the gods that spring would eventually return, mistletoe carried the most significance. Mistletoe is, by any standards, one of nature's phenomena; a hemi-parasite which uses the host plant as a growing platform, colonising soft-barked deciduous trees, especially apple, but also hawthorn, blackthorn, lime, poplar, maple, willow, plum, rowan, crab apple and occasionally oak trees. Mistletoe growing on oaks in the sacred oak groves of the Druids was the 'Golden Bough' of Celtic legend and held in great reverence by them. One can imagine the impact on superstitious early man, believing all plant life had ceased, suddenly seeing a clump of growth with green leaves and yellow berries glowing like a golden mist in a shaft of sunlight, high in the bare, lifeless branches of an oak tree. A startling confirmation that the ritual bonfires and sacrifices were doing their stuff; that spring really would come again, bringing warmth, fruitfulness and new life.

Mistletoe and its association with fertility featured prominently at the Roman festival of Saturnalia honouring Saturn, the deity of agriculture, when all the accepted codes of conduct were reversed; men dressed as women and masters as servants. There were pageants, uninhibited banqueting, bonfires and abandoned dancing; houses and streets were decorated with mistletoe, holly, ivy or other evergreens, and '*strenae*', twigs of laurel, holly or mistletoe with sweetmeats attached, were given as gifts.

In Norse mythology, the son of Odin, Baldur the Beautiful, was killed by an arrow made from mistletoe and his grieving mother Freya, goddess of love and beauty, banished the plant to the tops of trees. When Baldur came back to life, Freya made mistletoe a symbol of peace and love, hence the tradition among Norse and Germanic Celts that enemies meeting under mistletoe would accept a truce until the following day and this custom is believed to be the origin of kissing under the mistletoe. In the early Middle Ages, kissing boughs made from two bisecting wooden hoops woven with holly, ivy, laurel and rosemary, with a bunch of mistletoe suspended from them and hung from a doorway became popular Christmas decorations, with kissing under the bough being part of the jolly Christmas gambols.

Curiously, where the Church tolerated churches decorated with evergreens at Christmas, mistletoe was anathematised because of its association to the Druids, except in York Minster. The descendants of Viking settlers clung to their pagan customs and for many centuries, a Mistletoe Service was held on the cathedral steps where York's miscreants sought absolution beneath a branch of mistletoe held aloft by a priest. A sprig of mistletoe still decorates the high altar during the twelve days as a reminder of ancient customs and the spirit of reconciliation, love and repentance.

Among the symbols of Christmas, none is more synonymous with the season of goodwill, peace and joy, than holly, with its shiny, dark-green spiky leaves and brilliant, blood-red berries. Holly is everywhere at Christmas; it is sung about in carols, is the principal illustration on thousands of Christmas cards; the centrepiece of wreaths; decorates homes and no flaming Christmas plum pudding is properly dressed without a sprig of holly berries browning in the flames. Since the 1850s, when the Victorians created Christmas as we know it today, tons of holly have been sold annually at the famous Tenbury Wells Holly and Mistletoe Auction.

The sight of a holly tree, standing lush, green and aglow with scarlet berries in the midst of midwinter barrenness, would have been a God-given emblem of hope. To add to the mystique, holly most commonly grows in the understorey of oak woodlands, where few plants can survive the overhang of a mature tree and the Druids believed that once the leaves fell from a sacred oak, its spirit moved to the holly growing nearby. The early Christian converts decorated their houses with holly and evergreens during the Saturnalia and Yule revels, as much out of previous habit as protection from persecution. The custom persisted as Christianity spread and holly became more Christianised than any other plant, with the Church claiming the red berries represented Christ's blood at the Crucifixion and the spiky leaves, the crown of thorns. This notion became so ingrained that holly became known as Christ's Thorn or Holy Tree, the name William Turner, the Elizabethan natural historian, gives in his *New Herball* of 1568.

No two plants have been more closely linked than holly and ivy. Song was an important part of pagan winter solstice festivals and these were to become semi-religious ballads – the origin of our Christmas carols – in which a quixotic relationship between the holly tree and the ivy plant appears to have been a popular topic. Ivy, with its black winter berries, was held in high esteem by early people; it was dedicated to Bacchus, the Roman god of drunkenness and its leaves, either bound round the brow or added to wine, were believed to prevent intoxication. It was an emblem of fidelity and for that reason, in ancient Greece, priests always presented newly-weds with a wreath of ivy after the marriage ceremony.

From the early Middle Ages, right up to the Interregnum when Christmas was banned between 1649 and 1660 by Cromwell's Puritan Parliament as 'a popish festival with no biblical justification,' any number of ballads, poems and love songs were written about the two plants. Henry VIII even wrote one alluding to his love for a *'lady true'* enduring, as holly and ivy retain their evergreen vibrancy despite the harshest winter. Singing contests between men and women were a common Christmas pastime, with men extolling the virtues of the 'masculine' qualities of holly, whilst women praised the 'feminine' qualities of ivy and one of the earliest examples from the fourteenth century is among the Harleian Collection of manuscripts in the British Library. Everyone's favourite Christmas carol, the *Holly and the Ivy*, undoubtedly dates from this period.

Happy Christmas and 'lang may yer lum reek'!

January

January and the depths of bleak midwinter: the gradual change in the weather cycle is more noticeable in this month than any other. Forty years ago in the Borders, a heavy fall of snow was inevitable, but now we can expect anything from snow to bitter iron-hard frost, driving rain, a freezing east wind that takes the skin off your face, or a westerly, warm enough to bring puzzled winter sleepers, stumbling blearily from their hibernacula – be warned, I once had a terrier bitten by an adder in the middle of the month. Regardless of the weather, I do love the stark beauty of January; the empty hedgerows, the bare, leafless trees, their naked branches reaching for grey, brooding winter skies; rooks flighting home to roost in the gloaming, geese calling under the moon, the heavy silence of woodland and the musty smell of decaying leaves.

There is an atmosphere of expectancy about January, particularly the early part, which has nothing to do with relief that the festivities are over, or one's own hopes and aspirations for the coming year. It is much more to do with nature's response to the magic of the winter solstice, after the dark anxiety of shortening days. The imperceptible lengthening of daylight creates a ripple of primitive elation, epitomised for me by the dramatic, eerie shrieks of a clicketting vixen I always listen for at this time of year, when I take the dogs for a late run before going to bed. When I hear her screaming that joyous, urgent summons across the valley, I see her in my mind's eye, running backwards and forwards in the moonlight, ears cocked, listening as the answering barks of the dog fox come ever closer. More than anything, this primeval mating call early in the first month of the year when winter still has a long way to go, is a vocal promise of eventual spring fecundity, regrowth and summer warmth.

I always think of January as the sportsman's month and have never understood the post-Christmas exodus of fair-weather friends to Caribbean sun or alpine snow – there is little more than three weeks left of the season, why waste the best of it? The hours of daylight are still short and aficionados will be making the most of what is left: gone are the champagne at elevenses and a leisurely lunch, those in the know will be ready to move off sharp at nine o'clock and shooting through until the blackbirds start chinking and then eating – guns and beaters all together – in some dog-friendly pub or farmhouse kitchen.

Wildlife is much more wary now, which has as much to do with their sensing an imperceptible change in the season, as having their habitat disturbed over the previous three months. Before the infamous Blair Bill of 2004 criminalised hunting with hounds, a travelling dog fox in January provided the most terrific sport and even now, trail hunting within the law can provide wonderful days and the pleasure, simply of being out with horse and hound at that time of year. Despite the vagaries of politics or the weather, there is no such thing as a bad day's hunting. Pheasants are stronger, higher, wilder, faster and better eating than those shot before Christmas; for an unpredictable and sporting driven bird, a high, wide, long crossing January scorcher, is very hard to beat – unless you happen to be the walking gun. Early in the season when the leaf is still on the trees, a walking gun is not a particularly enviable place to be – one can get some spectacular snap shooting at the occasional nonconformist seen briefly between trees as it ducks back over the beating line, but vision is severely restricted by foliage.

It is entirely different to the magic of walking through with the beaters under bare January woodland. You can watch the whole drive unfolding, the beaters' dogs fossicking backwards and forwards, the birds driven forwards and the performance of guns in the line. Rabbits scuttling for cover, jays yammering in protest, perhaps a woodcock jinxing back and pheasants poised on one leg, heads cocked deciding which way danger lies, slinking under bracken or brambles and erupting out when flushed. This is when the old, long-spurred rebels who know the form, defying stops and beaters come swinging back above the leafless canopy, higher than those going forward, giving a walking gun the chance of a truly memorable shot.

If January is the sportsman's month, it is also the month of the next generation and all the more important for that. The 'smart' days are over and now is the time the bright-eyed young are given the opportunity of standing on a peg with their father beside them and experience their first driven day. It is a sight that warms the hearts of keepers, beaters and guns alike, and should one of them connect, the congratulations and applause after the drive is something they will never forget. January is a slack month in farming and the fever of business or bureaucracy is mercifully slow to recover from the Christmas break, for once one has that precious commodity, time and there is no better use of it than to further the education of the young. Walking up hedgerows together, to see what the terriers can find, or bracken banks on the moorland fringes in the hope of a rabbit or two and maybe, a woodcock. Sitting beside a flight pond in the gloaming as teal hiss by, or standing on the edge of a wood waiting for pigeons to flight in to roost and listening to the wind moaning in the branches. It is incomparable shared experiences like these, when one can pass our knowledge on and the true meaning of shooting and conservation, best illustrated by King George VI when he remarked: 'The wildlife of today is not ours to dispose of as we please. We have it in trust. We must account for it to those who come after.'

When I look back over decades of red-letter January days, an estate in South West Scotland on the banks of one of the Solway estuaries, where my teenage son and I were invited for many years, sticks out in the memory more than any others. The 7,000-acre estate was predominantly salt marsh, wetland, bog and grassland, with some arable and 150 acres of scattered small woodland plantations, rising to about 650 metres at the highest point. It was a wonderful wild bird shoot which had scarcely changed since shooting records began and like all wild bird shoots it was entirely weather dependent. On a wet stormy day when the glass was high, I have seen Canada geese, whooper swans and greylags all rising from the same pond; teal circling like pigeons and so many wisps of snipe in the adjoining bog they looked from a distance, like clouds of midges; the following day the wind dropped, temperature plummeted and there was hardly any movement. Traditionally, it was only shot in January and at this time of year the walkers, birdwatchers and cyclists had all migrated elsewhere, giving a wonderful feeling of having nature to ourselves. The diversity of landscape provided an incredible variety of quarry and a bag of ten different species during the day was not unusual – the record is twelve. Among them, one might expect teal, mallard, pintail, widgeon, goldeneye, gadwall, shoveller, common pochard, snipe, woodcock, pheasant, pigeon, rabbits, greylag, Canada and pink-foot geese.

A day would start long before daylight and if it was blowing a westerly gale likely to bring geese down into range, the keeper would lead us across the creeks and drains of the salt marsh to the merse edge, where we would hunker down in the reeds and wait for the distant roar of thousands of geese lifting off their shore roosts with the dawn. Then the breathtaking sight and sound of skein after skein of pink-feet calling to each other as they poured up the estuary, splitting off towards their favoured inland grazings and if the keeper had his flight paths right, giving us the chance of a shot. If there was no wind and geese were expected to be high, we would drive to the foot of the ridge running through the estate and carry bags of decoys up to one of the grazing fields at the top. Lying up behind a stone dyke and watching the first light glinting on the mudflats and water of the estuary below, and the armada of geese coming ever closer; skeins passing overhead and then a pack peeling away to circle the decoys and the keeper blowing his goose call to draw them in.

Breakfast was followed by stalking geese feeding on the fields, scuttling bent double along hedgerows, to kneel in position as the keeper crept round to the other side and the whoosh and clatter of wings as he put them over us. Then we would walk up a succession of the little woodlands for pheasant, rabbits and the occasional woodcock, or one of the many bogs for snipe, the odd mallard and those wild marsh pheasants, so utterly unlike reared birds, that explode out of rushes and rocket straight upwards. At some point during the morning, we would hide in the reeds beside one of the flight ponds and wait for the keeper to push the duck off. Lunch was a sandwich quickly eaten in the shelter of a farm building, before moving on to a different part of the estate and more woodland, bogs and rough ground on the edge of the salt marshes, with the endless cacophony of great flocks of waterfowl rising and falling out on the estuary, until pigeon started flighting in to roost. Pigeons are always tremendous sport and there was a very productive roost in old deciduous woodland at one end of the estate, but at the other, a platform had been built in the top of the trees in a spruce plantation, where the pigeons come at you like driven grouse. As light began to fade, the day ended with a duck flight; widgeon whistling all around you, the rapid wing beats of teal, the clumsy splash of a mallard landing and, in the distance, the glorious clamour of geese settling back on their shore roosts. You won't hear that in Musha Cay or Courchevel.